# AGING: THE PARADOX OF LIFE

**Robin Holliday** is a Fellow of the Royal Society of London, and the Australian and Indian Academies of Science. He has worked in several different biological fields. The first was the repair and genetic recombination of genes, chromosomes and DNA molecules. He devised an important DNA intermediate in genetic recombination, now known as the "Holliday structure" (or Holliday junction). He later initiated extensive research on the mechanisms of cellular ageing of normal human cells, and how they differ from immortal cancer cells.

This interest in ageing broadened to other biological systems and in 1995 he published the book *Understanding Ageing.* He was also a pioneer in the field of epigenetics, which is the study of the mechanisms for the unfolding of the genetic programme for development. Apart from the basic DNA code, there is another way heritable information is superimposed on DNA, and this has very important effects on the control of gene expression.

He obtained his Ph.D. at the University of Cambridge, England, and carried out his research at the John Innes Institute, Hertford, England, the National Institute for Medical Research, Mill Hill, London (where he was Head of the Genetics Division), and finally at CSIRO laboratories in Sydney, Australia. He has travelled extensively to international scientific conferences in the USA, Japan, Canada. Australia, India, and many European countries. He has himself organised such conferences in France, Italy and the UK. He has published over 250 scientific papers and several books.

By the same author

The Science of Human Progress
Genes, Proteins and Cellular Ageing
Understanding Ageing
Slaves and Saviours
Origins and Outcomes

# AGING: THE PARADOX OF LIFE

## OF LIFE

### Why We Age

Robin Holliday

 Springer

MT

A C.I.P. Catalogue record for this book is available from the Library of Congress.

ISBN 978-1-4020-5640-6 (HB)
ISBN 978-1-4020-5641-3 (e-book)

Published by Springer,
P.O. Box 17, 3300 AA Dordrecht, The Netherlands.
*www.springer.com*

*Printed on acid-free paper*

2/1/08

# Contents

# Preface

At the end of the 20th century a remarkable scientific discovery emerged. It was not a single discovery in the usual sense, because it was based on a series of important interconnected insights over quite a long period of time. These insights made it possible for the very first time to understand the biological reasons for ageing in animals and man. This book explains what these reasons are in non-technical language.

For centuries people have been puzzled by the inevitability of human ageing. It has often been referred to as a mystery, or an unsolved biological problem. Indeed, the famous zoologist Peter Medawar, who was to become a Nobel prizewinner in 1960, delivered in 1951 an important lecture - his inaugural lecture after being appointed Professor of Zoology at University College, London. The title was "**An unsolved problem in biology**," The unsolved problem was ageing, and when it was published the following year it had a strong strong influence on the scientific study of ageing. The German zoologist, August Weismann, in the late nineteenth century had suggested that ageing was essential so that generation could follow generation, and this allowed Darwinian evolution to occur. Medawar showed that Weismann's reasoning was false, because he explained – for the first time – that animals in their natural environments die mainly from predation, disease and starvation, and rarely reach natural old age. Such ageing is seen only in protected environments, for example, in animals which are domesticated, or kept in zoos, where they are well fed and looked after. Darwin also realised that many more offspring are produced that can survive to adulthood, and amongst these offspring there is competition in a hostile environment, which results in survival of the fittest.

There had been many scientific studies of ageing before Medawar, and many more were to follow. Much of this work could be called descriptive, for example many comparisons were made between young and old animals. Innumerable differences were documented, but these were often hard to interpret. Many experimental systems have been and still are being used. These include nematode roundworms, fruit flies, mice and rats, and even yeast and fungi. There were many theories of ageing that seemed to be competing with each other. A particular proponent of a theory often claimed it could explain everything. One of the most puzzling features of the ageing of mammals was the fact that similar changes occurred in old animals, but at very different rates in separate species. For example, mice and rats become old or senescent after about three years of adult life. Domestic dogs and cats do so around the middle of their second decade. Humans show body changes comparable to old rodents, cats and dogs after 70-80 years, although we all know that a minority reach a lifespan of 100 years.

Until quite recently this was not understood. For most of the second half of the twentieth century ageing remained something of a biological mystery. Thousands of scientific papers were published, many long reviews and several weighty books. Yet they came to no firm conclusions about the biological reasons for ageing, or why animals age at different rates. It was no accident that three books with positive affirmative titles were published in the 1990s. The first was by Leonard Hayflick "**How and Why We Age**" (1994); then my own "**Understanding Ageing**" (1995), and finally Steven Austad's "**Why We Age**" (1997). In addition, there was a book written in more popular style: Tom Kirkwood's "**Time of Our Lives**" (1999). These books are by no means the same, but they come to a similar conclusion, namely, that *ageing is no longer an unsolved problem in biology*.

In science, some discoveries immediately make their mark. A very good example of that was the momentous discovery of the structure of DNA by Jim Watson and Francis Crick, which earned them the Nobel Prize in 1962. Other discoveries are ignored, or even ridiculed. Gregor Mendel's breeding experiments with peas resulted in the elucidation of the laws of inheritance, which were published in 1865. This was ignored and then "re-discovered" at the beginning of the 20th century. The German scientist Alfred Wegener proposed the theory of continental drift in 1915, and based it on very sound arguments. Most geophysicists, geologists and geographers dismissed it out of hand. Yet today it is accepted.

It can already be said that the many observations and insights that explain ageing will not be accepted as established knowledge for a long time. The field is still full of scientists, and non-scientists, who are just happy to go on speculating about the "mystery" of ageing. As I have just said, many propose new theories of ageing, as if *their* theory will by itself explain everything. In fact, of the various theories of ageing that have been proposed over the years, several undoubtedly have a degree of truth. What we need now is refinement of existing ideas, not entirely different ones. Also, there are many clinicians and scientists who talk about "anti-ageing medicine". Many seem to regard ageing as a disease that can be cured like any other. I will discuss the myth of excessive prolongation of life in Chapter 9. The myths are largely due to complete ignorance of the reasons for ageing. The aim of this book is to dispel ignorance.

# Acknowledgements

My interest in the biology of ageing began in the 1960s. Many thanks are due to all those colleagues who in one way or another, by experiment or discussion, contributed to my present knowledge and interpretations of the field. They are too many to list here, but their names will be found in my earlier book *Understanding Ageing*. More recently, I have appreciated many discussions with Leonard Hayflick and our agreement that ageing is no longer an unsolved problem of biology. I also thank Jenny Young who typed many early draft chapters, and my daughter Mira Holliday for her help with the figures and final compilation of the manuscript.

# Author's Note

This book is largely written in non-technical language, with very few scientific references. These references are to be found in my earlier book *Understanding Ageing*, which was written for scientific readers. Some use of scientific and medical terminology is unavoidable, and I have provided at the end a glossary explaining or defining particular words. The words ageing and senescence are used more or less interchangeably. At various times 'the process of ageing' is referred to, and at other times 'the processes of ageing'. Both are legitimate, but the latter is more scientifically correct. Gerontology is the scientific study of all aspects of ageing, including social, demographic, and so on. Biogerontology is the study of the biology of ageing. Geriatricians are physicians with expertise in the care of old people.

# Chapter 1. Longevity

Today, almost every society contains old individuals, and children soon learn, through language and family, that they have elderly relatives, who one day will die. This was not always the case, because when humans first found themselves in natural environments, lifespans were very much shorter. The major causes of death were disease, attack by predators and shortage of food or water. A human population can sustain itself if infant mortality is about 25 per cent (as it is today in the great apes in their natural environments) and annual mortality is about 7 per cent thereafter. Under these circumstances the expectation of life at birth is less than 20 years. Females who reached reproductive maturity would expect to live to about 28 years and bear on average about 6 children. These figures are similar for one of the most primitive tribes in existence today, namely, the Yanomami Indians that inhabit a forest region of South America. In such societies, aged people are not very common, and death from old age is certainly the exception rather than the rule.

In Western societies with good health care the expectation of life at birth is now about four times higher than in primitive ones: more than 80 years for females, and just a few years shorter for males. There is not much infant mortality, and most subsequent mortality in the following decades is due to accidents, suicide, homicide, or occasional intrinsic disease, such as cancer. For a population, the shape of the survival curve is like that shown in Figure 1A. It is clear that the force of mortality starts to increase only quite late in life. With constant annual mortality, the survival curve is 'exponential' as shown in Figure 1B. Through history a major factor that increased longevity was the development of agriculture, which ensured a much more reliable supply of food. Even so, civilisations as advanced as the ancient Romans and Greeks had survival curves that were not very different from exponential, as shown in Figure 2. There was high infant mortality and expectation of life at birth was only about 25 years, and at one year was about 34 years. Infectious disease was the main cause of death, and it was not until most of this was eliminated in the twentieth century that the expectation of life rose very substantially. This was due to several major advances: the first was the discovery that disease was caused by infectious agents, usually bacteria or viruses. Then hygeine was greatly improved so that the chances of infection

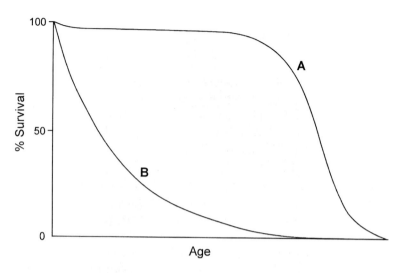

*Figure 1.* **A**, the survival of individuals in developed countries with good health care. In the early decades, death is mainly due to accidents (including homicide). Later the force of mortality increases, and few reach the age of 100 years or more. **B**, an exponential survival curve, which occurs when mortality is constant with time. Thus, after a given interval of time, 50% of individuals are alive, then after the same interval, 25% are alive, and after the same interval 12.5%, and so on

were greatly reduced. Finally, immunisation and antibiotics eliminated much of infection which still occurred. In addition to that, the quality of health care gradually improved through the twentieth century in developed countries, and even in many undeveloped ones. It is not uncommon for individuals a third world country to have an expectation of life at birth of 60 years.

Many inanimate objects have exponential survival survival curves. A simple example is a collection of wine glasses in a restaurant, as there is a constant probability that they will be broken at any time. Their survival is just the same as Figure 1B. One can extend the example to plastic glasses that do not break when they are dropped on the floor. Instead they accumulate scratches, and when enough of these occur, they are likely to be discarded. Their survival time is therefore more like Figure 1A.

There are in fact two measures of longevity. The one I have so far discussed is statistical and applies to populations of individuals, as shown in Figure 1. The other measure is the maximum lifespan. Amongst any population there will always be one that lives the longest. In Western societies with accurate records of births, the documenting

*Figure 2.* The poet W.H. Auden as a young man (top), and two years before his death in 1973 (below)

of longevities achieved is completely reliable. The oldest known individual was Jeanne Calment, who lived in France and died at the age of 122 years. Many reports of much longer lifespans will be discussed later in Chapter 9 and none have been authenticated. It is a fairly obvious generalisation that the larger the population that is properly documented, the greater is the maximum lifespan. This is an important point because many have made strong statements about the maximum lifespan of other animals based on rather small numbers of individuals in zoos. For example, it is often said that the maximum lifespan of gorillas or chimpanzees is very much lower than human lifespan, at around 50 or 60 years. Yet this is based on perhaps half a dozen animals kept in zoos, whereas human records include millions of individuals. If a few humans were kept in captivity, how long would their longevity be? Perhaps 70–80 years, which is 40 years less than Madame Calment. Although modern medical practice has greatly increased expectation of life at birth, it has not had much effect on maximum life span. Certainly there are far more centenarians than there used to be, but there is not much evidence that maximum lifespan is changing. There is no one alive today who is as old as was Madame Calment. One day there will be, but it will be one individual amongst many millions who will die younger.

Identical twins have the same genes as each other, but they do not have identical lifespans. They are more similar than that of non-identical twins or siblings of the same sex, so genes obviously have a role in determining lifespan. However, the differences between identical twins indicate that chance events are also important. Such events might be environmental or they may be intrinsic. The shape of the survival curve in Figure 1A is what would be expected if multiple events result in the increase in mortality at later ages. Thus, events can be occurring throughout life, but only when a sufficient number of them occur are the effects of ageing seen. (The lifespan of plastic glasses in a restaurant were mentioned earlier; they suffer multiple scratches and are then discarded). This very general interpretation is confirmed by experiments with animals which are genetically identical. When male and female mice are inbred for many generations, they end up with the same set of genes. These are known as inbred lines, and they can be exploited in all experiments in which the influence of different genes must be eliminated. When the longevity of a population of inbred mice is determined, it is found to be similar to Figure 1A, but with a maximum lifespan of about three years. As in the case of identical twins, there must be chance or random events which are important in determining longevity. Old mice and old people have quite similar changes in their cells, tissues and organs, but humans have a lifespan which is about thirty times longer than mice. This used to be a major puzzle in the field of ageing research, but now scientists know why this difference exists, and why other animals have their own characteristic longevities.

Everyone understands the effects of ageing, and we have a rough idea how old people are from their appearance. It may therefore come as a surprise that scientists have some difficulty in actually determining the age of single individuals. There is no individual adult feature that can be examined which is an accurate indicator of age. There are many features which taken together give an overall estimate, perhaps within a few years. This is again what might be expected if multiple events occur, and at least some of them are subject to chance. Nevertheless the effects of ageing are extremely obvious. Figure 2 shows photographs the poet W.H. Auden as a young man, and two years before died. They show that time has enormous effects on our facial features.

Another feature of human longevity is very important. Why are babies born young? The answer is that the germ cells, the sperm and eggs, do not age in the way that cells in other parts of the body do. There is plenty of evidence that human parental age has very little, if any, effect on the longevity of their progeny. There is a significant increase in some kinds of mutations in sperm as males become older, and older women have an increased risk of chromosomal abnormality in their progeny, such as Down's syndrome, which is caused by an additional small chromosome Nevertheless, there is absolutely no evidence that the children of older parents are physically or mentally inferior to those from young parents, nor that they have a shorter lifespan.

It can be said that the germ cells enjoy immortality, but very few of them are actually transmitted from one generation to the next. It is therefore better to say that they are potentially immortal; if they were not, populations of animals would not survive, but instead become extinct. The differences between germ cells and body cells with finite survival time will be discussed in several different contexts in later Chapters.

It used to be thought that *Homo sapiens* was the longest lived mammalian species, but this might not be true. Recently it was discovered that an adult bowhead whale had a harpoon embedded in its body which was of a type not used for 200 years. So this whale may have been at least that age. This is remarkable, and needs to be corroborated, but 200 years is still only a minute fraction of evolutionary time. The same can be said of giant tortoises, which live longer than humans. There are good biological reasons why a few species, including some plants, may have long life spans, and it should not be thought that it demands special explanation.

Although so much happens and so much can be achieved in one human lifetime, it can also be said that our lifespan is brief in terms of human history. Civilisations arose about 8000 years ago, and since then there have been about 350 generations of human life. Hominids walking upright appeared on our planet perhaps half a million years ago, a period comprising 20,000 generations or so. Anyone who lives to 60 years has experienced about one hundredth of one percent of human history. Why does this situation exist? Having developed to an adult, lived adult lives and produced children, why do we then disappear from the ongoing population? Why do different mammalian species have such different longevities? Why do most animals that breed quickly have short lifespans, and those that breed slowly have long lifepans? Why do human beings live as long as they do? Why is the average lifespan longer in women than men? Scientists now know the answers to these questions, and they will be found in this book.

# Chapter 2. Body Architecture

The focus of this book is the ageing of human beings, but much of what is presented applies to other mammalian species with different lifespans. The ageing of birds is in many ways similar to mammals, but as the evolutionary relationships decrease, then more differences from humans are seen. In this Chapter, I review the evolved structure and physiological design of the human body, with particular reference to the many non-renewable structures.

Animal cells, unlike bacteria and plants, have no rigid wall, but only an outer membrane. This facilitates chemical communication between them and allows the whole organism to function as a unit. The cell enclosed by the membrane is a highly complicated functional unit. It has a nucleus within which are the chromosomes. Human cells have 46 chromosomes (except sperm and eggs which have 23). The chromosomes contain the genetical material, or DNA. This is an enormously long thin molecule and the genetic information is encoded in a linear array of just four functional units, abbreviated to A, T, G and C. The DNA in every cell has a total length of about 2 metres (adding together the 46 chromosomes) and the number of units is about 6 billion ($6 \times 10^9$). This DNA is tightly packaged in the chromosomes, because the nucleus itself is only a few microns in diameter (one micron is one thousandth of a millimetre). The complete sequencing of the four units revealed that there are about 60,000 genes in the DNA of each cell (30,000 from each parent), and these code for the structure of proteins. Proteins are made up of a defined sequence of amino acids, of which there are just 20 kinds. The linear sequence of amino acids folds up in a particular way to form a specific functional unit, so that each protein has its own defined structure, and also a specific function. Most proteins are outside the nucleus, and held inside the cell by the membrane. Proteins are the very stuff of life. Many are catalysts or enzymes, which carry out all the metabolic reactions essential for living organisms. The membrane itself contains many proteins, tightly associated with lipids. (Lipids are the components of fat). The membrane determines what comes into and what goes out of cells, and it can transmit information to, or receive information from other cells. The total number of cells in a human being is enormous. In the brain alone there are about a billion cells ($10^{12}$), and in the body as a whole around 30 times as many. All these cells result from the division of a single cell, the fertilized egg. As development proceeds, through embryo, foetus and child to adult, the cells follow various pathways which leads to their specific differentiation into the many cell types found in different tissues and organs. Many will never divide again through adult life, others keep dividing continuously, such as those which renew skin, or the lining of the gut. In many tissues, and muscle is a good example, the cells cease division, but can nevertheless be replaced by a pool of quiescent cells which can be stimulated to divide and differentiate into new muscle cells. Thus, the muscles essential for locomotion, and other movements, are capable of some regeneration and repair. But in the heart, very little if any repair is possible, because there is no similar pool of quiescent cells. Heart muscle cells also have a different structure from those in the other muscles. The heart is a pump and its ongoing function is entirely dependent on the activity of all, or a large proportion, of its cells, very few of which can be replaced.

This raises a crucial question. How long can an individual cell be expected to survive? There are cells and structures in the living world which can survive a very long time, such as bacterial spores or many plant seeds. These are in a highly dehydrated state, and they essentially have no respiration or metabolic activity. Cells in an active animal are very different as they contain a high proportion of water and they also depend on energy from respiration. Under these conditions, innumerable chemical reactions are occurring. Many of these comprise the wide range of activities characteristic of any living cell, but some provide background 'noise' which can interfere with those normal activities. In particular, respiration generates oxygen free radicals (often abbreviated to ROS - reactive oxygen species). These are short-lived but highly reactive. During their very short lifetime they can damage DNA, proteins, or lipids in the cell membrane. As we shall see later, the cell has defenses against these free radicals, but these are not perfect. It is known, for example, that the genetic material DNA is continually bombarded by free radicals and suffers damage to its individual components. Most of this damage can be repaired, but not all of it. The DNA is a double helix, consisting of two long strands coiled around each other. A break in one strand can be repaired, but breaks in both strands close together causes the molecule to fall apart, to fully break, and this damage is much harder to repair.

The adult brain consists largely of non-dividing cells called neurons which are entirely dependent on a good oxygen supply, and are very active in chemical metabolism. These cells are therefore subject to damage from free radicals throughout their life. The DNA is never replaced, only repaired, and the repair is not perfect. This means that sooner or later a cell will suffer damage which may end its useful life. That is not to say that neurons commonly die from damage to DNA, because there can also be many abnormal changes in proteins. Although many such molecules are broken down and replaced, others resist such turnover and slowly accumulate. These abnormal molecules can then form insoluble aggregates, which strongly interferes with normal neuronal function, and may eventually kill the cell. In addition, proteins may be partially degraded to components called peptides, some of which are known to be harmful. Thus, for several reasons, neurons are not immortal. They have finite survival time, so as we get older and older the number of neurons gradually decline. In Alzheimer's disease and other dementias, the rate of cell death is accelerated, and much of this appears to be due to the accumulation of insoluble proteins or fragments of proteins. There are cells in the brain which are capable of cell division, but the cells responsible for brain function, the neurons, cannot divide. The inability of brain tissue to repair or regenerate itself is described in a well known text book of pathology as follows:

**Localisation of function makes the brain inherently vulnerable to focal lesions that in other organs might go unnoticed or produce only trivial symptoms . . . . . . This vulnerability of the brain to small lesions is compounded by its very limited capacity to reconstitute damaged tissue. There has clearly been a serious error by the celestial committee in its design of an organ that is vital for biological survival and yet is both the most vulnerable to, and least tolerant of, focal damage.**

The biological fact is clear: an active living cell incapable of division can never be immortal, and if it cannot be replaced by other cells, then complex structures such as the brain will eventually age. The brain must last a lifetime, and therefore some of the longest living non-dividing cells we know are those in the brain and nerves of the longest-lived animals such as giant tortoises, or the whale mentioned in Chapter 1.

The heart is similar to the brain in that it consists of non-dividing cells, entirely dependent on a supply of oxygen, and very active in chemical metabolism. In this case, a large proportion of the available energy is devoted to rhythmic muscle contractions. For the animal to survive, muscle cells have to function without interruption throughout life. The heart is a highly efficient pump, but since it cannot effectively repair itself, or replace lost cells, it necessarily has a finite lifespan. Loss of heart function is the major cause of death in old age.

The major blood vessels consist of connective tissue, muscle cells and an essential inner lining consisting of a single layer of cells. All these components are subject to changes during ageing, and unfortunately most of these cannot be reversed or repaired. One reason for this is that essential proteins, such as collagen and elastin in the connective tissue components, are laid down early in life and cannot be replaced. They are structural components subject to 'wear and tear' including chemical cross-linking between different molecules. These lead to a loss of the elasticity that is an essential feature of normal arterial function. It is also known that other proteins in the major arteries suffer from other protein damage, such as oxidation or glycation (the binding of a carbohydrate such as glucose). As well as this, the layer of cells lining the artery is subject to damage, particularly atherosclerosis, which through the build up of atherosclerotic plaques can eventually block the blood flow. Although minor arteries and capillaries can be replaced (for example, when a wound heals), the major arteries must function continuously and their anatomical design is such that they cannot be effectively repaired, except by medical intervention.

With regard to ageing, teeth are a very instructive component of the body. After milk teeth are discarded, the larger adult teeth replace them. These teeth are subject to wear and tear, and also to bacterial attack. Although they contain nerves and a blood supply, they are essentially mechanical structures which cannot be expected to last forever. In the case of humans, the existence of dentists allow them to last much longer than they would normally do; alternatively they can be replaced with artificial teeth. This is a clear case of intervention in the normal ageing process, which I will discuss more fully later on. Teeth show very clearly that the distinction which is often made between "programmed" ageing, and ageing through wear and tear, is quite artificial. Our genes determine the size, strength and structure of teeth. That is clearly part of the programme for development. The teeth, however, are subject to wear and tear, so gradually lost their normal structures. Clearly the lifetime of teeth is determined both by the genetic programme and the extent of wear and tear. In a sense, teeth are "designed to last a life time," which is of course the result of a long process of biological evolution. Much the same conclusion applies to other non-replaceable components of the body, such as the vascular system and brain.

The eye is another vital body component which cannot be expected to last indefinitely. The lens is made up of specialised proteins known as crystallins. These are laid down early in life, and are never replaced. As we all know, their particular property is transparency, but the constituent proteins are subject to a variety of abnormal chemical changes that simply accumulate with time. One change which is essentially continuous is the loss of elasticity of the lens, this leads to a gradual decline in the ability to change the focus of the lens. (This is known as loss of accommodation). Eventually transparency itself can be lost leading to the formation of cataracts. The probability of this happening varies between individuals, and can be accelerated by environmental influences (for example, by the wood smoke generated during cooking in third world countries). Lenses are subject to age-related changes simply because there is no mechanism to replace their protein molecules. The retina contains a layer of cells, each of which has photoreceptor structures. These cells, like neurons, can never be replaced. In the rod cells, the photoreceptors look like a pile of pennies. Each component has a finite functional lifetime, which is very short relative to the lifetime of the cell. So new photoreceptors are added to the top of the pile, nearest the light source, and old ones are removed from the bottom. During a lifetime each retina cell may turn over about three million photoreceptor elements. The problem is to get rid of the old ones that are not required. Cells underneath the light-sensitive layer of the retina continually take up and degrade the used-up photoreceptors, but eventually insoluble proteins start to accumulate. With time the retina loses normal structure and function, and blindness can follow. Again, the structure of the retina allows it to last a lifetime, but it is not a steady-state renewable structure, so cannot be expected to last forever. There are other non-renewable structures in the body, such as the filtration structures in the kidneys, and bone joints. Their gradual erosion can lead to osterarthritis and inflammation. The sound sensitive components of the ear are also non-renewable, and hearing often declines with ageing.

What is striking is that all those changes which bring about a decline in structure or function, occur with a broad degree of synchrony. This makes biological sense, because one cannot imagine the evolution of organisms over millions of years in which some components could last a very long time, whilst other vital components had a much shorter lifespan. It is said that each year for a dog is the equivalent of seven for a human. This is broadly true, and if you examine an elderly

12 year old dog, you will see a similar range of age-related tissue and organ changes as you would see in an 80 year old human. There are very good reasons why different animals have very different lifespans, which I will return to later on.

What about renewable structures such as skin? The epidermal cells of the skin are continually dividing to produce cells that synthesise keratin. These skin cells differentiate themselves to death by turning into the tiny parcels of keratin which comprise the outer layer of the skin. The tough protective keratin is continually worn away during normal life, and simply replaced by continued division and differentiation of skin cells. Yet we all know that skin ages, indeed, it is one of the best indicators of a person's age. Part of the reason is that the layer underneath the epidermis, the dermis, is not renewable in the same way. Also, it contains the protein collagen which is very long-lived and becomes cross-linked as we age. This reduces its elasticity, and also the degree of wrinkling of the skin as a whole.

It has become clear from many experimental studies that dividing human cells cannot multiply indefinitely. Many types of dividing cells have been grown in the laboratory (using so-called tissue culture techniques) and it has been shown in every case that the cells eventually become senescent and cease growth. This was first demonstrated from cells called fibroblasts, which are present in the dermis and many other parts of the body. They synthesise collagen which is the most abundant protein in the body, and a essential constituent of all connective tissues. Fibroblasts grown from the skin of young individuals divide significantly more times than those from old individuals. This indicates that the cells are dividing slowly during our lifetime, but are not endowed with the ability to divide forever. Thus, tissues which appear to be renewable are not in a steady state, although in this case they can probably fulfil their major functions for more than a lifetime.

The ends of chromosomes have a unique DNA structure known as the telomere. When a linear DNA molecule replicates, or duplicates, a short stretch at the end of the molecule would be lost, but the telomere prevents this happening and the length is maintained. However, this special mechanism also depends on an enzyme called telomerase. It turns out that many cells of the body that can only divide a given number of times lack the enzyme telomerase. As they divide their telomeric DNA gets shorter. Other cells, known as stem cells, have telomerase so can divide indefinitely. A popular theory of cellular ageing is based on the supposition that the progressive loss of the ends of chromosomes has a number of adverse effects which result in senescence and the inability to divide. It is important to understand that the telomere theory of ageing is not applicable to the many body cells that never divide.

When skin and the underlying tissue is wounded, the processes of repair come into operation. This includes the division of fibroblasts and other cells, the synthesis of collagen, and the renewal of a vascular supply. Wound repair, as well as blood clotting, is an essential mechanism for animal's ongoing survival. It is possible that the additional reserves of cell division that we see in fibroblasts, and some other types of cell, is an insurance against excessive tissue damage, which might otherwise be fatal. Although the repair of wounds is generally efficient, it has its limits. Major wounds leave abnormal scar tissue, often with inadequate blood and nerve supply. When major nerves are severed they cannot regenerate. Although broken bones will rejoin, the loss of part of a limb, such as a finger, cannot be replaced by regeneration. Some vertebrate species, such as amphibians, can regenerate lost parts quite effectively, but this ability was lost during mammalian evolution.

---

Some non-renewable structures

---

~ Brain neurons
~ Heart muscle cells
~ The major arteries
~ The lens of the eye
~ The retina of the eye
~ Sound sensitive ear structures
~ Kidney filtration structures (glomeruli)
~ Bone joints
~ Adult teeth

---

A vast amount of study and research on the anatomy and physiology of human beings and other mammals, clearly demonstrate that the design and architecture of the body is incompatible with indefinite survival. What kind of complex animal might survive indefinitely? Can one conceive of a design which would never age? Probably one could, but it would not at all be like the human and animal bodies we know. For example, there would not be one major vascular system, but two. This would enable one to function, whilst the other is shut down and either effectively repaired, or completely replaced by regeneration. This rejuvenated system could take over, and allow the other one to be repaired or replaced later on. Such duplicated system would also depend on a pool of cells capable of indefinite cell division, unlike the cells we actually have. The central nervous system provides a particular problem. An immortal animal would have to have the means of replacing senescent neurons, or the cells of the retina. We can imagine that learning is compatible with formation of new cells (as seems to occur when birds learn new songs), but it is much harder to envisage the replacement of memory. There is every reason to believe that human memory is dependent on long-lasting structures, including neural circuits. To imagine the transmission of the information which comprises memory from one set of neurons to a new set, is indeed mind-boggling! To put it bluntly, a dynamic or plastic central nervous system, with full renewal of component parts, is scarcely compatible with the sophisticated functions of the human brain. Indeed, it could be said that ageing is a price we pay for having language and all the other unique features of the human brain.

# Chapter 3. Maintenance of the Body

The fertilised egg develops to a foetus, a baby, a child and finally an adult. It is obvious that this sequence of development is extremely complex. However, there is not only the unfolding of the genetic programme for development, but there is also a set of mechanisms that can recognise and repair any mistakes or defects that arise. These mechansims also maintain the adult in an active normal state for several decades, so that reproduction and the raising of offspring can proceed. Ageing is due to the eventual failure of this maintenance. Many defects continually arise in molecules, cells and tissues, and most of these are repaired or reversed whilst we are fit and active. Eventually, however, maintenance is unable to cope and the deleterious changes progressively accumulate to produce all those changes that we see during ageing. In this Chapter I briefly descibe the most important maintenance mechanisms that keep the body in a functional state for so long. In fact, the long human lifespan – longer that other land mammals – is due to more efficient maintenance and repair, and I will return to this theme later on. Research on maintenance mechanisms comprises a considerable proportion of biological research, but most of these scientists do even realise that what they are doing has anything at all to do with the processes of ageing.

In the previous chapter, I mentioned the repair of wounds, which is a very obvious form of body maintenance. But there are many other maintenance mechanisms that are continuous, and also far less apparent to us. One of the most important is the repair of DNA. This structure contains all the genetic information essential for the preservation of the life of an organism, and also for the reproduction of the same organism. To understand the central role of DNA in the biology of the organism, a few more words about its structure and properties are necessary. DNA contains all the genetic information in two complementary copies, wound round each other to form the double helix. The four chemical units or components in DNA, A, T, G and C, are arranged in the double structure as follows: A is always opposite T, and G is always opposite C. This means that the two copies are complementary: the sequence of units on one strand defines the sequence on the other. In human DNA there are about 6 billion units of information (A-T and G-C) pairs, present in every cell, except red blood cells which have no nucleus. DNA is a chemically stable molecule, but quite a few of the component units are altered every day of our lives in every cell. These are chemical alterations are thought to be due to "spontaneous" damage in DNA. Spontaneous damage can occur as an unpredictable, uncommon chemical event, which may result in the loss of one of the units of DNA (simply a gap in the sequence of instructions), an abnormal chemical change in a unit, which in some cases may give it a different coding property. However, it is likely that most "spontaneous" DNA damage has a cause.

The power-houses in each cell are small organelles which produce energy from respiration. This process depends of course on a continual supply of oxygen, and an energy source such as glucose. During respiration some short-lived reactive components are produced as a by-product. These are known as oxygen free radicals, or ROS. These are neither molecules nor atoms, in the usual sense, and they have a transient existence. In fact, their survival ends when they react with some cellular component, that can be DNA, protein or lipid. The reaction often produces an abnormal or defective product, for example, an oxidized amino acid in a protein, or a peroxide in a lipid component. In the case of DNA, the reaction changes a single normal component to one which is abnormal. Because there are so many DNA components per cell, even the alteration of a very small fraction can amount to several thousand defects in every cell everyday of our lives.

It is not surprising that organisms have evolved defence mechanisms against ROS. There are important enzymes that degrade them even more quickly than they can damage important component of cells. Also, there are chemicals in cells and in blood serum which are known as antioxidants, and examples are vitamin E, vitamin C and carotenoids. These are present in our diet, and many take additional supplements, because they believe, or have been told, that they get rid of dangerous oxygen free radicals and are thereby beneficial. Be that as it may, it is undoubtedly true that the defence against ROS is a very important maintenance mechanism.

In spite of this defence, DNA still suffers damage, although scientists argue about the proportion of damage which is due to reactive oxygen species, and the damage due to other causes. What is important is that essentially all this damage is repaired by proteins. These are repair enzymes which can almost instantly detect any abnormality in DNA. The abnormality is cut out by a process known as excision repair, and the short tract of one strand of DNA units which is removed is replaced by normal undamaged units, as is shown in Figure 3. This is possible because DNA has two copies of the genetic information. All

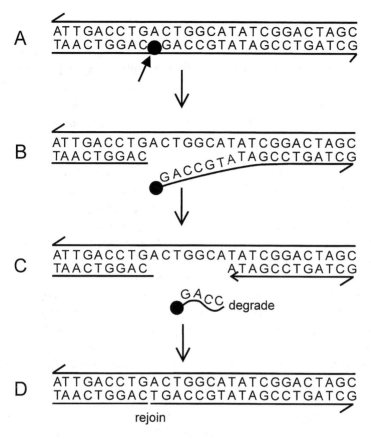

*Figure 3.* The excision repair of damage in DNA. The damage – a filled circle (A) – is recognised by an enzyme that cuts one strand and strips away a short portion of the molecule (B). This fragment is released and degraded (C). The single stranded gap is filled by another enzyme, and the final small gap is rejoined (D)

this is carried out not by one enzyme, but by three or four different ones. Indeed, from studies of bacteria and other organisms, it is now well known that if one repair mechanism fails, for whatever reason, there are back-up mechanisms that can also spot and repair the DNA damage. Thus, there are several pathways of repair, each depending on a small set of enzymes. The whole comprises a complex and remarkably efficient means of keeping our DNA intact. This is dramatically shown when an individual inherits a defect in just one of the many genes necessary to produce all the necessary enzymes. Such an individual can show dramatic abnormalities. One of the earliest studied was a defect in the repair of damage in DNA caused by the ultra-violet light in sunlight. Such an individual soon develops multiple skin abnormalities in all parts of the skin exposed to light. These abnormalities affect pigmentation, the production of keratin, and cell division which can lead to the development of cancer. At least six genes are known which if mutated to an inactive form, produce a similar range of symptoms. Many other genes are now known that are necessary for the repair of DNA damage, other than that produced by UV light.

There are other related systems of repair. When a cell divides, every gene replicates to form two new copies. This process is extraordinarily accurate. It is known that one mistake occurs in about one hundred million, or a billion units of DNA. (In contrast, a skilled typist is likely to make one mistake in about 5000 letters). This accuracy depends in fact on proof-reading mechanisms built into the machinery of DNA replication. The instructions for making two DNA molecules from one is in the sequence of units in the DNA itself. The replication is carried out by a complex of enzymes and accessory proteins. When a mistake occurs, it is normally recognised: the wrong unit is cut out and thrown away, and the correct one is inserted. However, this can also fail and a defect is then present in the newly-formed stretch of DNA. There is another proof-reading process which can recognise this defective component, cut it out, and replace it with the correct component. If this too fails, for whatever reason, then the defect remains in DNA and becomes a mutation. This is fixed or made permanent in the DNA after one more cell division. A mutation is like the typist's error which is not recognised during proof-reading and persists into the final print-out of the text.

DNA does not make proteins directly. Instead the information in each gene is read into a transcript, known as RNA. After this has been processed into a form known as messenger RNA, it leaves the nucleus with all the instructions necessary for the synthesis of a particular protein. The synthesis of RNA and proteins is also accurate, but much less so than DNA. There are also proof-reading devices that can detect errors in the synthesis of RNA from DNA and the synthesis of the amino acid chain of a protein from RNA. There are good reasons why these molecules are less carefully monitored than DNA, and there are also consequences as well. As we shall see later on, errors or defects in DNA, RNA and proteins may be an important component of ageing. It should be noted that a mutation in DNA (ie. the change of one component unit to another) will change the RNA message and that very often leads to the change of one amino acid for another in a protein. Errors can occur either in the formation of the RNA message, or in the reading of the message, and these also result in amino acid changes. However, there is a big difference between a mutation and an error in the transfer of information. Whereas a gene mutation causes all the proteins coded from by the gene to be altered, errors in the synthesis of RNA or protein affect only a minority of protein molecules and the others are quite normal. So mutations can have a more profound effect, but they are also far less common than errors in the formation of RNA or proteins.

Most damage in proteins does not come from errors in synthesis, but from other chemical changes that occur during the proteins life-time in the cell or organism. These are often most clearly seen in long-lived proteins. The crystallin proteins of the eye lens have already been mentioned. These are laid down during the formation of the lens, and never replaced. With time abnormalities accumulate. These include oxidation (by ROS), the cross-linking of different protein molecules, spontaneous changes in amino acids, or the attachment of sugars to proteins. Alternatively, the correct folding of the amino acids chain may change to produce what is known as a denatured, or partially denatured, molecule. Most amino acids can exist in two forms, one left handed (L) and one right handed (D). In all living organisms only the L forms are found in proteins. (This is a quirk of nature to do with the origin of life, and its subsequent evolution). However, there is a given low probability that an L amino acid will change spontaneously into a D amino acid. This is known to occur in long lived proteins in teeth, and also in the lens, but at a very slow rate. Collagen is also a long-lived protein in the body. In this case the commonest abnormality is the cross-linking of different molecules which thereby changes their physical properties.

Only a few types of proteins are very long-lived, because most of them (especially enzymes that are active in metabolism) have a relatively short existence. Why is this so? There are several reasons, but one which is very important is the need to recognise and get rid of defective molecules. Enzymes are highly specific in the chemical reactions they catalyse. If they become altered (for example, by the type of damage that occurs in lens crystallin), they may lose their normal specificity, or more commonly, they may simply become non-functional. If nothing was done about this, all these molecules would begin to clog-up every cell in our body. To maintain normal organisation and efficiency it is essential to get rid of them. So there are special mechanisms which recognise abnormal protein molecules and degrade them to their individual amino acids. These mechanisms depend on enzymes that can chew up proteins, known as proteases. Such enzymes are an essential part of our digestive system, but within cells their role is much more subtle and discerning. They have to recognise abnormal molecules, and leave normal ones intact. This ability to discern abnormal and normal is only partly understood, but there is no doubt that it is an efficient process in all cells. The turnover of proteins occurs continuously during normal metabolism; many amino acids are re-utilised in the synthesis of new protein molecules, but some are degraded and end up as urea. The continual excretion of urine is the major consequence of protein turnover, and this turnover is an essential maintenance mechanism to keep our cells in a normal functional state. No process is perfect, however, and as we shall see, the failure to degrade abnormal protein molecules has serious consequences.

Microorganisms grow and divide rapidly, within a time scale of minutes and hours, whereas animals live a long time and reproduce over a time scale of months or years. This could have very serious consequences for the animal Pathogens and parasites are always seeking new environments, that is, new hosts in which they can breed. They can, in principle, easily outstrip the growth of their host and kill it. To prevent this happening animals have developed a very important maintenance mechanism, known as the immune system. The critical feature of this system is the recognition of anything foreign in the body. This "foreignness" is usually an invading pathogen or parasite. It can be a bacterium or fungus, an animal parasite or a virus. Many are specialised invaders; others can live and reproduce outside animals, as well as within them. Viruses are not themselves complete living organisms, because they depend on the machinery of a living cell to reproduce. Most of the viruses that infect us can only grow in particular kinds of animal cells, either in the laboratory or in the living animal. Pathogens and parasites can infect us through wounds, in the respiratory tract, or in food, but some are transmitted by other animals, such as mosquitoes or other insects. The variety of all this danger in the environment needs no elaboration here. Nor is it possible to elaborate on the complexities of the immune system. Most of the cells of the immune system are in blood, and they arise in the bone marrow, but there are other specialised organs too, notably the

thymus and spleen. As is usually the case, there are alternative mechanisms of defence; if one fails, there is normally a back up mechanism.
Foreignness is often a protein, or part of a protein, which is not present
in the animal itself. One way to combat it, is to synthesise a protein
antibody that can inactivate the foreign protein. Alternatively, a host
cell containing a virus or other intruder can be recognised and killed by
cells of the immune system. In this way the pathogen is also destroyed.
Some carbohydrates, for example particular components of bacterial
cell walls, can evoke antibody responses, which can inactivate or kill
the invading microorganism. Toxins produced by bacteria can also be
recognised as foreign and destroyed.

One feature of the immune system which was originally puzzling to
immunologists, is the ability to reject tissues from another individual
of the same species. The rejection of skin grafts or organ transplants
is very well known, but obviously in a natural situation, such tissue
grafts never occur. It turns out that the same ability to distinguish self
from non-self is again operating. Cells from all individuals contain
particular families of proteins, which are highly variable in normal
population of animals. Thus, it is overwhelmingly probable that any
foreign cell has protein variants which differ from one's own cells.
During development of every individual, the immune system learns
which of all possible protein structures are self-structures; therefore
any other structure must be foreign. If turns out that the rejection of
proteins from another individual is just part and parcel of our bodies
ability to detect foreignness in invading pathogens. Needless to say,
the study of tissue rejection has played a very important part in our
understanding essential mechanisms of the normal immune system.

The crucial importance of the immune system is seen in infections
which destroy one or more components of the system itself. The best
known case is AIDS (acquired immuno-deficiency syndrome). After
infection with the HIV virus, there is a long incubation period, followed
by the progressive loss of normal immune defences. AIDS patients
commonly succumb to infections, or they may develop particular forms
of cancer. Cancer cells synthesise new proteins (or foetal proteins not
present in the adult) which are recognised as foreign. Thus, immunological surveillance is an important defence against cell abnormalities
in ones own body. All this seems rather far removed from ageing, and
immunology is a highly sophisticated science in its own right. Nevertheless, it is an essential maintenance mechanism, and if things go wrong

with the immune system, then deleterious consequences are seen. These consequences are one part of the ageing process, as we shall see.

The study of cancer is also a large part of all biomedical research. Why is the research so difficult? The answer, at one level, is quite simple. Cancer cells are obviously abnormal, but to understand their origin we have to first understand normality. That is, we have to understand how an organism develops from a fertilised egg into all the multiple components of the body, and also how all these multiple components are regulated and interact with each other. This is the normal situation, and it is highly complex. It is not fully understood, so how can one understand why something goes wrong when the normal situation is itself mysterious? What it amounts to is that cancer is a disease affecting normal cell maintenance. In the body, it is very striking that specialised cells – muscle cells, brain cells, cells of the immune system, skin cells and so on – are all very stable. Non-dividing cells keep their integrity for our whole life; most dividing cells produce daughters of the same type. However, during development of the organism and also in special stem cells in the adult, daughter cells may be different for their parents, because they acquire new properties, especially the properties of stable fully specialised cells. Cancer is, in effect, a disease of development, or a disease of stem cells. Something goes wrong, which means that the normal regulatory control, or controls, is lost. The cell becomes isolated from its neighbour, starts to divide, and continues to do so. Normal regulatory mechanisms are essential to prevent that happening. This regulation itself is a maintenance mechanism to prevent the body being taken over by rogue cells. There are genes called tumour suppressors which are part of the defences against cancer.

There is also quite a new area of study known as epigenetics. All cells (with some specialised exceptions) have the very same set of genes in their DNA, but the way genes are expressed in different cell types is not well understood. Obviously haemoglobin is made in red blood cells and not in brain cells. What are the globin genes doing in the brain? They are there, but not expressed, whereas specialised brain cell proteins are expressed, but not in blood cells. These epigenetic controls, and their stability, is now a very active field of research. It is well known that mutations in certain genes are liable to trigger cancer, and also that the incorrect turning on and off of particular control mechanisms, other than by mutations, is also important. It turns out that the stability of cells is greater in long lived mammals such as man, than in short lived mammals such as mice. This also provides new insights into the events of ageing.

Herbivorous animals obtain their energy from plants. These animals have specific advantages over plants, namely, their ability to move around to find new sources of food. They have also evolved innumerable devices to ingest and digest plant tissue. Some plants have evolved certain mechanical defences, but their main defence depends on chemistry. Most have developed the means of synthesising chemicals which are toxic to animals, so the latter learn to avoid eating them. Obviously all plants do not get eaten, nor do all animals die from plant toxins, so there is a dynamic balance between them. One part of this interaction is the ability of animals to detoxify harmful chemicals in their diet, which is a particularly important function of the liver. This organ produces a wide range of specific enzymes which can break down to harmless derivatives a huge variety of organic compounds synthesised by plants. This detoxification is an essential maintenance mechanism, since many plants in our normal diet contain chemicals which are routinely removed by the liver, or by other tissues. During evolution plants may come up with a new toxic chemical, but in turn the animal evolves an additional defence. This has been referred to as an arms race between species. (Human activity has of course, resulted in the production of thousands of completely new chemicals, some of which are released in the environment and some of which are toxic. Although the liver enzymes have never encountered such chemicals during evolution, they can nevertheless act on them in many cases. Unfortunately, such enzyme action may be imperfect and indeed harmful. For example, many hydrocarbon chemicals [containing only carbon and hydrogen] are carcinogens, but only because they are converted by liver enzymes to activated products containing oxygen, that are now known to damage DNA). However, for our purpose, the real biological situation is the relevant one, and this is the maintenance of the normal functional state of the animal. Normal detoxification is the sifting out of noxious chemicals in diet to produce harmless derivatives, some of which can be metabolized by other pathways. Although herbivores are most at risk, carnivores which feed on herbivores may also be at risk, so they have their own detoxification system. In some cases, and particularly in insects, animals have become resistant to plant toxins, and can store them for their own defence against predators. In other cases, particularly frogs and toads, animals have developed their own toxins as a defence mechanism against predators.

In the study of physiology, a very important theme is the variety of regulatory mechanisms which exist. An obvious example is the response to the expenditure of muscle energy – the intake of oxygen is increased. Also, the loss of blood, or an inadequate supply of oxygen at high altitude, is followed by an increase in the production of red blood cells. There are innumerable regulatory mechanisms involving hormones, or other chemical messengers, and many involve feedback loops. Physiologists use the word homeostasis to categorise all those regulatory mechanisms. The most important of all, which we almost take for granted, is temperature control. This helps provide a constant body environment, independent of outside influences. At first sight, it seems paradoxical that warm-blooded mammals and birds are so adaptable to changes in outside temperature. Many species, including man, can survive from the tropics to the arctic. It is the control of body temperature which makes this possible, whereas cold-blooded vertebrates are at the mercy of the temperature of their surroundings and therefore less adaptable. (It is even possible that this was a major reason for the extinction of the dinosaurs, and the selective survival of warm blooded animals). Temperature control and innumerable other regulatory mechanisms, comprise a vital set of body maintenance mechanisms. Homeostatic mechanisms are not perfect, and things can go wrong especially in adults beginning to age, with far-reaching effects on the body.

Cells respond to environmental insults by synthesising a defence mechanism consisting of a family of stress proteins. Since an increase in temperature was the first to be studied in detail, these proteins were labelled the "heat shock" family of proteins, but they also appear after other potentially harmful environmental effects. One consequence of increased temperature is the partial unfolding of many cellular proteins. Stress proteins are the same or similar to "chaperonins" which are necessary for the correct folding of many proteins. This can be regarded as a form of protein repair. Also, there are enzymes which can remove damage, such as oxidation, in several amino acids in proteins.

Another maintenance mechanism, which has only become fully apparent in the last few years, is a special process called apoptosis to get rid of unwanted cells. This is a cell suicide mechanism in which the DNA is fragmented, the cell shrinks, and is ingested by scavenging cells known as macrophages. It occurs during normal development of the organism, for example in the central nervous system, or the immune system, but a better known instance is the regression of the tadpoles tail as it changes into a frog. It turns out that this suicide mechanism is triggered more widely than was previously realised. It is the means a complex organism uses to get rid of cells which are damaged beyond repair. Their continued survival would place an unwanted burden on the organism, including in some cases the development of cancer.

Fat storage can be regarded as another important maintenance mechanism. In natural environments, the supply of food fluctuates. When it is abundant, some is stored in the form of fat. This is then used up in times of scarcity. Clearly this adaptation benefits survival in harsh or stressful environments. Unfortunately it has bad consequences in any environment where food is always plentiful. In this case fat keeps being added to the body and obesity results. This is an unusual example of a maintenance mechanism failing in an artificial man-made environment, whereas it is effective in a natural environment. Human obesity commonly leads to late-onset diabetes with all its physiological complications, and puts additional stress on the vascular system.

There is also maintenance of the outer surface of the body, by cleaning, washing and grooming. Much of this behaviour is to remove or keep at bay potential parasites, seeking entry via the skin. Particular grooming skills are seen in many primates, where hands are used, and there is frequently reciprocal grooming between related animals, which allows surveillance of those parts of the body inaccessible to the individual animals. It has recently been discovered that human skin produces antibiotics which kill bacteria on the surface of the body. As time goes on, more and more aspects of maintenance are revealed.

The most obvious maintenance mechanism, briefly discussed previously, is wound repair. Small wounds are fully repaired and leave no trace. Large wounds are repaired with the production of scar tissue. The clotting of blood is one part of the repair of tissue damage. Normal skeletal muscle can regenerate, if damaged, and broken bones can re-join. All these are necessary to maintain normal body functions. Nevertheless, in mammals repair and regeneration is a compromise. Lower vertebrates such as amphibians can often regenerate a severed limb, and many lizards regenerate tails. As evolution proceeded some of this regenerative ability was lost. In mammals and birds a lost limb is not regained by regeneration, and if major nerves are severed, they do not rejoin or regenerate. This is an important issue. It is clear that maintenance, as exemplified by wound repair, does not have endless resources. The extent of maintenance is a compromise between the investment in resources, and economising on resources. It is very likely that the balance between the two is very important in determining the lifespan of mammalian species.

---

Major maintenance mechanisms

---

~ DNA repair
~ Defences against oxygen free radicals
~ Proofreading in DNA, RNA and protein synthesis
~ Breakdown of abnormal proteins
~ The immune system
~ Detoxification of harmful chemicals
~ Wound healing, blood clotting, rejoining of broken bones
~ Cell suicide – apoptosis
~ Physiological homeostasis
~ Epigenetic stability
~ Stress proteins and protein repair
~ Fat storage
~ Grooming of fur or feathers

---

With regard to an understanding of ageing, the importance of maintenance and the eventual failure of maintenance cannot be stressed too strongly. A large part of biological research is devoted to a better understanding of maintenance mechanisms, but few of these scientists realise that their work, in one way or another, also relates to the study of ageing. It is also apparent that a very large number of genes are necessary for all maintenance functions, so these genes also have a role in determining longevity and lifespan. This is quite contrary to the view, put about by some who study ageing, that there are just a few "gerontogenes" which determine our lifespan, or that of animals. I will return to genes and ageing later on.

# Chapter 4. Multiple Causes

In Chapter 2, I described and listed a number of non-renewable structures in the body. Many of these can last a lifetime, but not indefinitely. What we see during ageing are progressive changes in these structures. Here I relate these changes to the multiple causes of ageing. The changes occur with a degree of synchrony, but sometimes one is in advance of the others, and then it is liable to be labelled a disease. Heart disease and Alzheimer's disease are good examples of this.

The heart is an efficient pump, but it consists of cells which never divide. As in the case of any mechanical device, with time deterioration sets in. Heart muscle cells may be lost, the valves, through constant use, can become calcified or otherwise defective. The major blood vessels are not easily repaired, especially their inner surface. Cumulative damage gives rise to the condition known as atherosclerosis. Cholesterol-rich lesions known as atheromatous plaques appear and gradually increase in size. These can eventually block blood flow, which in turn can lead to a failure of the normal oxygen supply to the heart, with a consequent heart attack, or lethal heart failure. The formation of blood clots or the detachment of plaques from the arterial wall can impede the blood supply to the brain. and cause a stroke. The arterial walls are elastic in young individuals, but with time the cross-linking of collagen and elastin reduces elasticity and thickening of the walls may also occur (hardening of the arteries). These changes often result in hypotension. This can affect kidney function, and high blood pressure can cause a stroke from a haemorrhage in the brain. It is frequently stated that "heart disease" is the commonest cause of death. But heart disease is the result of gradual changes over time, and is one component of ageing itself. It is labelled a disease because it can clearly occur in individuals who are otherwise quite healthy, whereas others may live to an advanced age with a strong heart and circulatory system.

The neurons of the brain cannot be renewed or replaced, and with time irreversible changes occur. These include the accumulation of defective proteins or protein fragments called peptides, and memory is thereby impaired. When this process occurs earlier than other ageing changes, it is labelled Akzheimer's disease, which at present is incurable. The incidence of Alzheimer's disease doubles every five years after the age of 65. Very elderly people who have most of the normal signs of ageing, may also have similar changes in their brain, but in these cases they are not usually regarded as victims of Alzheimer's disease. It is also established that neurons are lost in the brain with age, which will ultimately affect brain function. There are two major classes of protein in the body. There are those that continually turn over, that is, they are degraded and replaced with new molecules. Other proteins are very long lived, and this group includes the collagen and elastin, components of connective tissue, and the crystallin proteins in the lens of the eye. With time these long lived molecules accumulate changes such as crosslinking between different molecules, oxidation or glycation (the attachment of a sugar molecule). Individual amino acids in proteins can undergo a variety of chemical alterations which impair their function. All these chemical changes in long lived proteins are yet another cause of ageing.

Although enzymes called proteases can often recognise and remove altered proteins, they are much less efficient in removing aggregates of protein molecules. One such aggregate, together with lipid components, is known as the 'age pigment,' lipofuscin, which appears in many tissues in old people. Another type of insoluble aggregate are collectively known as AGEs (advanced glycation end-products). They are also widely distributed in ageing cells and tissues. Diabetics who have abnormal glucose levels, are particularly affected by the build up of AGEs. In Chapter 2 I referred to the problem of the turnover of photoreceptors in the light sensitive calls of the retina of the eye. Partial or complete loss of normal vision is a common component of ageing. Irreversible changes in the crystallin proteins of the lens leads to the formation of cataracts. A decline in hearing is also very common.

Mechanical structures in the body suffer from wear and tear. Limb joints are subject to continual mechanical stress. However excellent the biological design, eventually damage accumulates, which is often associated with arthritis. Lubrication and maintenance is very important, but it is not perfect, and eventually fails. For bone structure to stay in a steady state, there must be a balance between the formation and re-absorption of the bone matrix. This balance is lost in the late decades and bone strength declines, leading to osteoporosis and the increased likelihood of fractures.

The immune system consists of many different components, which must operate as an integrated unit. It is well known that the efficiency of immunity declines with age, probably because individual components are not making their normal contribution to the whole. Old people become more susceptible to infections such as influenza or pneumonia.

The immune system is capable of distinguishing self from non-self, and that is why grafts of tissue from other individuals are destroyed. Auto-immune diseases are due to a breakdown of the recognition of self cells or tissues. Obviously this has serious deleterious effects. Examples of autoimmunity increase with ageing, and a particularly painful example is osteoarthritis, or degenerative joint disease. However, osteoarthritis can also be caused by mechanical wear and tear of joints, which is very common in aged people. We should also note that the alterations in normal proteins during ageing may generate molecules that the immune system sees as non-self, and as a consequence there is an autoimmune reaction.

One example of efficient repair is the removal of damage in DNA in all the cells of our body (except red blood cells). Unfortunately this repair is not 100% effective. This is demonstrated when human cells are treated with a DNA damaging agent such as ultra-violet light or X-rays. The cells can survive low doses, but as the dose increases more and more cells are killed. One obvious reason for this is that the repair enzymes become saturated; they simply cannot cope with all the damage. But there are other reasons too. Sometimes two DNA lesions occur very close together, that are far more difficult to repair; and there may also be chemical changes that are not detected by repair enzymes. There is now plenty of evidence that so-called spontaneous damage in DNA is not completely repaired. It is known that gene mutations slowly accumulate in white blood cells with age, and also chromosome breaks, or changes in chromosome number. Such changes can occur both in cells that divide, or those that never do so. Since our genes control so many body functions, it is not surprising that DNA damage can ultimately have a multitude of different effects. Each defect or mutation may not be too serious, but as they accumulate their combined effects become greater.

The organelles that that produce energy from respiration are called mitochondria. They also contain one small circular DNA molecule. This is subject to change, and particulaly deletions of different parts of the molecule. This loss of genetic information obviously impairs mitochondrial function. Since there are many mitochondria in every cell, this may initially not have any serious effect. However, it is easy to see that as more and more molecules are affected, mitochondrial function as a whole becomes impaired. This can be regarded as a 'multiple hit' process, which can become an important component of ageing. It is thought that mitochondrial DNA damage is largely due to the formation of reactive oxygen radicals from respiration. Although these radicals are short lived, they can also damage other cellular components, such as cell membranes, proteins and the DNA in chromosomes.

Cancer is a major age-associated disease, apart from childhood leukaemias and some other fairly uncommon types of tumour. Epidemiologists have known for a long time that the rising incidence of the one of the commonest types of tumours (carcinomas) with age can be explained on the basis of several sequential events (in the range 4–6). Some have concluded that cancer is nothing to do with ageing; it is simply time-dependent! This becomes a semantic issue, because ageing is itself a time-dependent process, or more accurately, a set of processes In cancer, it is thought that the events must occur successively in a given cell lineage, because it is well known that tumours originate for a single precursor cell. The events could be mutations, and the chance of any individual cell acquiring all these genetic changes will be very small. Unfortunately, because there are so many cells in the body, the probability that at least one of them will eventually become cancerous becomes quite high. Of course, similar mutations may be occurring in all cells, but the vast majority do not reach the point where growth regulation is affected. Cancer is such an obvious body abnormality, that it is not surprising that it is labelled as a disease. Nevertheless, many old people who die from some other cause, commonly have several or many incipient tumours. As in the case of cataracts, heart disease and so on, if we all lived long enough, we would probably all develop cancer.

In recent years, it has become apparent that as well as mutations there are other important events that occur during the genesis of tumour cells. Epigenetics is the study of all those mechanisms necessary to unfold the genetic programme for development. This depends on DNA, but also information that is chemically superimposed on DNA, known as DNA modification or DNA methylation. As well as normal epigenetic events, there are also abnormal ones that can be referred to as epigenetic defects and these can be inherited from cell to cell. It is now known that epigenetic defects are very important in the development of tumours. Thus, one or more of the 4–6 events just mentioned could be epigenetic. It is also likely that these defects are an important component of ageing itself. The expression of genes in specialised cells is controlled by epigenetic mechanisms, although in most cases the exact way in which these operate is not fully understood. We would expect that epigenetic defects might result in the loss of gene expression, or perhaps equally important, the turning on of a gene which is not normally expressed in a given cell type. To give an extreme example, brain neurons do not have globin component of the haemoglobin in red blood cells. The expression of globin in a neuron could well have a bad effect on that cell. In Chapter 1, I mentioned that identical twins do not have the same lifespan, which shows that there random events are very important in determining the onset of ageing. Recently evidence has been obtained that identical twins have differences in DNA modification in their cells. These, by definition, are epigenetic changes, so this makes it all the more likely that epigenetic defects contribute to ageing.

One of the characteristics of the overall ageing process, or set of processes, is that a partial failure of one organ system can impinge on another. A good example of this is the increase in blood pressure that occurs during ageing. This may be largely due to the loss of elasticity of the walls of major arteries (hardening of the arteries). High blood pressure has deleterious effects, and may in the end lead to kidney failure, and the requirement of a dialysis machine, or a kidney transplant. High blood pressure can also result in the rupture of a brain artery, and a slight or a severe stroke may result. It can also have serious effects on the retina of the eye.

Hormones are chemical signals which are vital to sustain the normal balance between the functions of different organ systems. Thus, insulin controls the level of glucose in the blood. There are two types of diabetes, one of which can be of early onset, and is due to the loss of the cells that synthesise insulin. This type of diabetes can be controlled by the injection of the hormone, but these diabetics die about 15 years earlier than the average lifespan. The other type is called 'late-onset' because it becomes much more common in older people, especially if they are obese. This is due to abnormalities in insulin metabolism, for example, an insufficiency in synthesis, or just as likely, changes in the cells that normally respond to insulin. The cause is a matter for debate, but it is well known that this type of diabetes is age-related, and also that it has a variety of effects on other organs system, particularly the kidney, but also nerves and eyes, and an increase in atherosclerosis. This is an excellent example of one abnormality causing several others. Apart from diabetes, there are other less well known age-related hormone abnormalities. Therefore, we can add all these to the long list of established causes of ageing which is included in this Chapter.

Over the years many specific theories of ageing have been proposed. In fact, the history of this field of study is characterised by the overstatement of the importance of each theory. Thus, the proposer of a particular theory often claims that it explains the whole of ageing. What is striking that each theory can usually related to a particular cause of ageing, or to a failure of one of the maintenance mechanisms summarised in Chapter 3. For example, the somatic mutation theory proposes that changes in DNA are the cause of ageing. The mitochondial theory of ageing is even more specific in that is asserts that the loss of mitochondrial function is the cause of ageing. Various protein-based theories state that it is the irreversible accumulation of altered or abnormal proteins that are responsible for ageing. The free radical theory, which is one of the most popular, proposes that damage to DNA, proteins, membranes, and other cellular components, is the major cause of ageing. Another theory of ageing relates mainly to the many interactions between different parts of the body which are mediated by hormones. This is often called the neuro-endocrine theory of ageing, because many hormones are controlled by brain function. It is also to do with physiological mechanisms – often called homeostatic mechanisms – and the eventual failure of these to control normal body function. The immunological theory of ageing is based on the decline in immune function and the increase in autoimmune reactions. Each of these theories, and several others I have not mentioned, specify a particular cellular component, or a particular type of tissue or cell. Since many of these cellular components, cells or tissues, undoubtedly deteriorate during ageing, we can be confident that many of the theories have some degree of truth. No theory is entirely correct, in that it alone 'explains' ageing, but several have a considerable degree of validity. Thus we need to have a more global view of all theories. because only in this way can we explain the multiple and irreversible causes of ageing.

Many of the causes of ageing I have outlined take place in non-dividing cells. What about cells capable of cell division? In these cases cells that die or are lost, for whatever reason, should be be replaced by new cells. However, it is now well established that many types of dividing cells have in fact limited growth potential. After dividing a given number of times, they themselves become senescent and fail to divide again. This is often referred to as the 'Hayflick limit,' because it was first discovered by the American scientist Leonard Hayflick, using human connective tissue cells known as fibroblasts. These and several other types of cell are simply unable to maintain themselves. However, some cells in the body, and particularly stem cells, may have the ability to divide indefinitely. The question at issue is whether cells that have limited proliferation contribute to ageing of the body. The short answer is that we do not know, but there is some evidence that it does, and especially from experiments with animals. Cells taken from old animals divide significantly fewer times than those from very young animals. This is shown in Figure 4. The youngest cells divide about thirty times, and the oldest about nine times. These observations indicate that the number of cell divisions possible at the outset, get partially used up during a lifetime. It is quite possible that dividing cells have the potential for renewal and replacement, but with some divisions in reserve. It would nevertheless be surprising if the Hayflick limit in

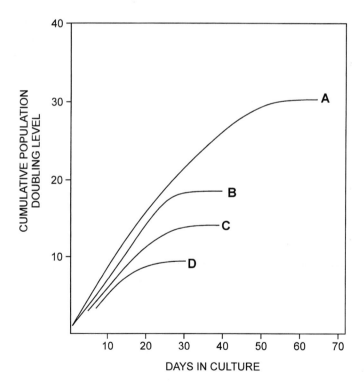

*Figure 4.* The growth in culture of skin cells from Syrian hamsters of increasing age. **A**, cells at 13 days gestation; **B**, 3 days neonatal; **C**, 6 month old adult, and **D** a 24 month old adult. Cumulative population doublings are broadly equivalent to cell divisions

some kinds of cell had no discernable effect on ageing. The immune system depends on cell division, and this may be one reason for its decline during ageing. We also know that the repair of external wounds is slower is significantly slower in old people compared to young ones. Probably there are changes in internal tissues or organs, that are corrected or repaired less efficiently during ageing, and therefore contribute to general impairment.

It is striking, however, that some of the most obvious manifestations of ageing do not contribute to the spectrum of changes that eventually cause our demise. One example of this is the greying and whitening of hair, or the loss of hair. This in no way can be regarded as a pathological change. Similarly, the wrinkling of skin is a major characteristics of ageing, yet by itself it does not cause much harm. It is due to progressive changes in the dermis and epidermis, and particularly the loss of elasticity. It is a symptom or manifestation of ageing, but not a cause. Muscle strength declines with age, which due to a loss of weight and volume of muscle tissue. This can be said to have indirect effects on the well-being of old people. If movement and balance are impaired, then falls may follow, with broken hips or other damage. We all know that some old people may not recover from such accidents.

Medical intervention counteracts some causes of ageing, and the best example is care of teeth. Teeth wear out with age, and they get infected, but these problems can be dealt with by dentists. Examination of Egyptian mummies has shown that tooth abcesses were common, and in some cases lead to septicaemia which would usually have been lethal. With modern medical treatmentent, defective joints in the hip can be replaced; kidneys, livers or hearts can be transplanted; hearing aids alleviate deafness; cataracts can be removed and replaced with an implant, and spectacles have been used for centuries to improve vision. New treatments for delibitating diseases in the elderly can be expected, but we certainly should not glibly assume that the fundamental causes of ageing will be prevented or reversed. The myths of life extension will be explored in a later chapter.

The fact that there are multiple causes of ageing also means that the successful treatment of one age-associated pathology, will be almost certainly followed by the emergence of another. The design or architecture of the body is not compatible with indefinite survival, in spite of the fact that we have many efficient maintenance mechanisms. Maintenance is essential for normal development to the adult and for the several decades of adult life. After that things start to go wrong, as we are no longer able to sustain the status quo. Eventually the complexity and order that characterises life itself is lost, and death follows. On our death certificate a single cause of death may be given, but the reality is that if we did not die from one cause, we would die from another. The fact that one organ-system fails first, should not blind us to the fact that there is a multiplicity of changes occurring in molecules, cells, tissue and organs during ageing.

---

Some causes of ageing

---

~ Progressive decline in heart function
~ Irreversible changes in major arteries
~ Brain damage and loss of neurons
~ Protein abnormalities (e.g. crosslinking, oxidation, glycation)
~ Accumulation of insoluble protein aggregates
~ Mutations in chromosomal DNA; chromosomal changes
~ Deletions in mitochondrial DNA
~ Hormonal and physiological imbalance
~ Abnormalities in gene function leading to cancer
~ Abnormalities in the modification of DNA (epigenetic defects)
~ Partial or complete loss of hearing and vision

---

As I have said, the multiple causes of ageing mean that there is some truth in many, or all, of the serious theories of ageing. The various molecules and cellular defects come from several or many causes, and it is highly likely that these causes relate in one way or another to the individual theories which have been proposed. In the previous chapter, the various mechanisms of body maintenance were discussed, and it is very striking that there is also an almost a one-to-one relationship between each maintenance mechanism and each theory. In other words, the theory supposes that a particular maintenance function eventually breaks down (eg. failure of DNA repair leading to DNA damage; loss of immune function leading to auto-immune reactions and so on). Since all maintenance mechanisms are important, it follows that all serious theories of ageing are important too. We need to take a more global overview of these theories in understanding the multiple causes of ageing. This global overview is that ageing is due to the eventual failure of maintenance. Maintenance is essential for normal development to the adult and for the several decades of adult life. After that things start to go wrong, as we are no longer able to sustain the status quo. Eventually the complexity and order that characterises life itself is lost, and death follows. On our death certificate a single cause of death may be given, but the reality is that if we did not die from one cause, we would die from another. The fact that one organ-system fails first, should not blind us to the fact that there is a multiplicity of changes occurring in molecules, cells, tissue and organs during ageing.

# Chapter 5. The Ancient Origins
# of Ageing

Not all organism age. It is well established that populations of bacteria or yeast can go on growing forever, as long as nutrients are provided. Many plants reproduce vegetatively (or asexually), by forming tubers, bulbs or corms, or they can be propogated by cuttings. Thus, the continuous growth of complex living organisms such as plants is perfectly possible. Why then do so few animals have this ability?

The interrelationship between sexual reproduction and the existence of ageing have been known since the last century. At that time the German zoologist August Weismann realised that there was a clear distinction between germ-line cells and the cells of the rest of the body, often referred to as somatic cells. When an embryo is developing, a particular group of cells is set aside. These are the germ-line cells which will eventually give rise to the sperm or the eggs of the adult. Since these cells produce the next generation of individuals, and each following generation, they are potentially immortal and do not age. We say that germ cells are only potentially immortal, because clearly most are not immortal. In fact, only a minute fraction of male sperm and a small fraction of female eggs, or egg precursors called oocytes, will survive. Moreover, individual lineages die out if offspring do not reach adulthood and reproduce. More significantly, whole species can become extinct, either from natural causes or by human intervention, and then both their germ cells and somatic cells are lost forever.

The crucial issue is the intrinsic difference between germ cells and somatic cells, which make the former potentially immortal and the latter mortal. How did this situation come about, and why did it come about? To answer the question we need to understand the early origins of organisms with many cells. No one doubts that the earliest organisms were single free-living cells that divide by simple division, two cells from one, four cells from two, and so on. The resources of the environment are not unlimited, so sooner or later cell division slows down or stops. Many of the cells may die, perhaps from lack of nutrients and of water, but others survive. They are potentially immortal, like germ cells. These cells have single nuclei containing the genetic material DNA. Nature explores all avenues, and one was the evolution of organisms consisting of groups of cells. Another was the evolution of elongated cells containing many nuclei, like many fungi and algae which exist today. A further evolutionary step was the appearance of cells with specialised functions, such as spores which can resist dessication, UV light or heat, or combinations of motile and non-motile cells. Most importantly, these organisms developed sexual reproduction in which cells with a double set of chromosomes alternate with cells with a single set. This depended on the evolution of a special form of cell division (known as meiosis) in cells with two sets of chromosomes, in which there is an exact halving of each set to produce nuclei with one set of chromosomes, either in sperm and eggs, or in many cases in resistant spores. The full complement of chromosomes is restored by the fusion of two such nuclei, or cells containing them. The early evolution of sex, and the maintenance of sex during subsequent evolution, is a hotly debated topic in innumerable articles and books. It is likely that cells with two sets of chromosomes, and therefore with two copies of each gene, allowed organisms to store much more genetic variability, since the two genes of each type could have slight, but significant, differences in their DNA. The cell division which produces nuclei with only one copy of each gene, also re-assorts different genes into a multitude of different combinations. Thus, all the progeny in sexual reproduction are genetically different from each other, whereas in asexual reproduction they are all the same, forming a clone. For our purpose, we can just accept the fact that sexual reproduction evolved, and also that it is beneficial because it is so widespread in nature.

There is not much doubt that early multicellular organisms reproduced both by sexual reproduction and also by asexual, or vegetative reproduction. Today we see many species, such as fungi and algae, which reproduce by both means, and the same is true of innumerable plants. Most plants produce flowers, pollen, eggs and seed, but many of these can also reproduce asexually. Here we are concerned mainly with animal evolution. Simple animals today reproduce both sexually and asexually. This is seen most clearly in the hydras and corals which can bud off new individuals or polyps. Colonies of animals can form, in which it may be hard to determine whether an individual animal is a single polyp or a whole cluster of polyps. Other animals, obviously related to primitive forms, are known as flat-worms. Many are free living, and others have become parasites of contemporary animal species. They have no body cavity, unlike all more advanced animal species, and they retain remarkable powers of regeneration.

Some will produce buds that develop into whole animals, and it has been shown experimentally that pieces of an adult animal can regenerate the whole organism again. This regeneration is due to special cells in the body which are totipotent, that is, they can form all the other types of cells found in an adult animal (much like stem cells in the embryos of complex animals) These animals are in some way more like plants than more advanced types of animals. They have potentially immortal germ cells and also some immortal body cells.

Given the existence of both sexual and asexual reproduction in early animals such as these, what are the evolutionary choices determining subsequent events? We do not have to go very far up the evolutionary ladder to find a very different situation. Simple free living round worms, or nematodes, are found in vast numbers in ordinary soil, feeding on bacteria, or other microorganisms. Unlike flatworms, they have a body cavity containing the gut, and they also have a mouth, anus, muscles, nerve cells and so on. The remarkable feature of these animals is that their development culminates in an adult with a constant number of somatic cells. Moreover, these cells never divide, and the only cells capable of division in the adult are the germ cells that will form the next generation. In discussing the non-dividing cells of our brain, or our heart, I earlier made the fairly dogmatic assertion that these cells cannot be expected to survive forever. The same applies to the simple roundworm. The non-dividing cells of the adult cannot be expected to survive for ever, and indeed they do not. An individual worm will only live for days or weeks, even if it is in an ideal environment with plenty of food. These animals clearly exhibit ageing, a fact which has been seized by many experimental biologists in attempting to document of what actually happens during this ageing process, and also what determines the actual lifespan under good conditions in a laboratory.

The details of these changes during the ageing of nematodes need not concern us here. The important thing is that clear-cut ageing has evolved in this small animal. During its evolution it has dispensed with asexual reproduction and relied entirely on its germ cells, either in male, female or hermaphrodite individuals. How could this be advantageous? This brings us back to the fact that in normal environments animals usually die because they are eaten by predators, killed by pathogens, or they run out of food or water. This means the population is age-structured: there are many more young adults than old ones. In fact, in a normal population there are few or no ageing individuals. Under these circumstances the best strategy for survival is to reproduce.

Consider two animals **A** and **B**. **A** can reproduce sexually, and also
has a body which can maintain itself indefinitely, by replacement of
somatic cells as required. **B** can only reproduce sexually, and its body
consists of non-dividing cells which will age and die. In an ideal
environment **A** might do better, because it can survive and reproduce
indefinitely, where **B** can only reproduce for a limited period of time.
In Darwinian terms **A** appears to be fitter than **B**. Therefore, one might
think, ageing could not evolve. The paradox is that **B** is fitter than
**A** in a real environment. In this real environment, lifespan is short
because of the innumerable hazards. **B** reproduces as quickly as it can,
whereas **A** is using body resources to preserve itself. This investment
of resources is simply wasted, if the lifespan of the animal is ended by
predators, starvation or some other lethal event. The resources saved
by **B** are channelled into reproduction instead. *This means that **B** has
more reproductive success than **A**, and natural selection will favour
its survival.*

The same argument arises from the fundamental reasoning of
Malthus and Darwin. Malthus realised that the full reproductive
potential of a species such as man can never be realised. Such reproduction, if many offspring survive and also reproduce, is logarithmic
or exponential. To put this in simple numbers, suppose each family
consisted of four children, each of which also had four children, and so
on. Then the number of individuals would double in each generation.
In 10 generations there would be $2^{10}$ or 2048 individuals arising from
two, and in 30 generations there would be $8.6 \times 10^9$ individuals (i.e. 8.6
billion individuals, which is more than the present human population).
In natural environments that could never happen, because resources
are never unlimited. In most cases a hazardous environment keeps the
numbers of individuals at a more-or-less constant level, albeit with
fluctuations due to the amount of food available or to variation in
environmental hazards. (Man has in fact broken this general rule by
manipulating the environment to suit his own life style and reproduction, hence the enormous number of individuals alive today. This
is itself relevant to the evolution of human longevity, as we shall see
in a later chapter).

Darwin fully realised the significance of Malthus' conclusion. Most
animals and plants produce far more offspring than can possibly
survive to maturity. Thus, there is a struggle for survival between
individuals of the same species, and also competition between species.
There is also genetic variation between individuals, and natural

selection will act to select the fittest variants. Fitness is a measure of the ability to survive and reproduce. All individuals are subjected to environmental hazards, and obviously there is a strong element of chance in survival. Nevertheless, the genes also influence the animal's chance of survival, and therefore there will be natural selection of genes. All this has been outlined in recent years with considerable persuasion and clarity by Richard Dawkins in his book *The Selfish Gene*, and also by many other authors.

Darwin wrote books not only about evolution, but also about many other biological topics. Curiously, he never seemed to discuss ageing. It is curious, because it is just those features of natural populations which he seized or in this theory of natural selection, that also explains the evolution of ageing in animal populations. I previously described the roundworm as a simple animal, which it certainly is, but it is also highly specialised in its lifestyle. The important point, however, is that all types of animals find themselves in hostile environments, which produce age-structured populations. For this reason, ageing probably did not evolve once, but several times in different taxonomic groups. One of the most successful of these comprises the vast array of insect species. Many of these species, like the roundworm, have adult bodies with very few, if any, cells capable of division, apart from their germ cells. In general, they have limited powers of regeneration, and their bodies have finite survival times. This survival time in insects is extremely variable: from the few hours of the adult Mayfly, to lifespans of many months, or years in species which have a long period of dormancy in their life cycle. The determination of longevity of any one species is a complex issue, and in evolutionary terms, it is ultimately dependent on the ecological lifestyle of the species in question. However, in physiological terms the longevity of a species is dependent on the resources which are invested in maintaining the viability of the body.

The resources available to any animal can be broadly partitioned into three main areas. First, there are the fundamental features common to all living organisms, namely, feeding, digestion, respiration, metabolic activity to produce all the chemical components of cells, the elimination of waste as a by-product of metabolism, and waste in non-digestible food. Feeding also includes muscle activity and movement in the search for plants in herbivores, and for other animals in carnivores, or for hosts by parasites. Second, there are all those activities concerned with growth and reproduction: the development of the fertilised egg to the adult, the production of germ cells, sexual activity, and, in many cases, the care of offspring. Obviously, there is enormous variability in the reproductive strategy of various types of animals. Many fish, for example, invest most of their reproductive resources in the production of huge numbers of germ cells, whereas mammals and birds invest less in germ cells and much more in parental care, including the feeding of offspring. The third major investment of resources is in maintenance of the adult body, which, for mammals, was outlined in Chapter 3. The essential point is that the first allocation of resources is essential for all animals, but the second and third are highly variable. We can pinpoint a fundamental principle in saying that there is a trade-off between investment in reproduction and investment in maintenance. So we find animals which grow and reproduce very quickly but have short lifespans, and animals which grow and reproduce slowly with long lifespans. But we never see the other theoretical possibilities, namely, animals which grow and breed quickly with long lifespans; and animals which grow and breed slowly with short lifespans. The former is a physiological impossibility, and the latter would lead to rapid extinction.

---

Allocation of the resources available to a mammalian species

---

**Normal functions**
Biochemical synthesis
Metabolism
Respiration
Cell turnover
Movement
Feeding and digestion
Excretion

**Reproductive functions**                                              **Maintenance functions**
Gonads, gametes and sex                                                All those discussed and
Development                                                            listed in Chapter 3
Gestation
Suckling
Care of offspring

---

The above conclusions can be best applied within particular taxonomic groups, such as mammals, or birds. Amongst animals as a whole there is enormous diversity, and there exists for instance, large fish which produce vast numbers of eggs, and highly specialised insects, such as certain parasitic wasps, which may produce few offspring. We also see enormous diversity in the length of time an animal will breed. Some adults breed only once and then die. Investment in maintenance is essential up to the time of reproduction, but not thereafter. The classic case is the Atlantic salmon, which does not survive long after it has spawned. Other animals breed repeatedly over quite long periods. In these cases, the fertile adult body must be maintained in a normal healthy state for years or decades. However, there is no example of a complex animal in which the body is maintained indefinitely. Sooner or later ageing and death supercede reproduction.

It has sometimes been argued that some large fish or reptiles do not age in the usual sense. These long-lived animals usually increase in size throughout their lifespan, in other words, they have pools of dividing cells that keep growing and slowly increasing body weight. Nevertheless, the long survival time of a giant tortoise or large fish such as a whale shark, is still only a minute fraction of evolutionary time. It may well be that size itself is one determining factor of lifespan. The weight of an animal roughly increases as the cube of its length. So a large animal such as an elephant must have immensely strong limbs to walk or run. Given its basic anatomical design, there is simply a physical or mechanical limit to the size of any species, even the fish or whales that do not have to contend with gravitational forces, have other mechanical limits to their ultimate size. Thus any animal that continued to grow would eventually reach that limit. We see this also in trees; they keep growing, but eventually if they get too large and tall, they can no longer resist wind forces and fall over. Their anatomical design ensures the mortality of the individual, apart from those species which have special means of vegetative propagation, or have evolved additional mechanical support, such as the Banyan tree. Such a tree may be potentially immortal because it forms, in effect, a clone of identical individuals, radiating out from the centre. (There is a famous Banyan tree in the Calcutta botanical gardens which covers an immense area). No animal species more complex than a coral, has evolved any similar means of survival.

In conclusion, we can say that the evolution of different longevities is the result of animals' adaptation to a variety of environmental niches. The trend towards longer or shorter lifespans within different groups of animals must have occurred many times, and specific examples will be discussed later, including the evolution of human longevity.

# Chapter 6. Mice and Men

Although mammals have successfully adapted themselves to a wide variety of habitats and life styles, their basic physiological design is very uniform. Also, their mode of reproduction does not vary significantly. After a given period of gestation, offspring feed from their mothers milk, the mammary glands being the primary taxonomic feature which provides the name for the whole group. The variable features of mammals are, most obviously, their size and external morphology. This is the result of successful adaptation to habitats as diverse as the sea (whales and dolphins), the air (bats), or underground (moles). Yet all these animals retain constant anatomical features, for example, a given vertebra of a whale, in spite of the huge difference in size, has a very similar shape to one in a mouse, or in a human. The internal organs, vascular system, muscles and nervous system all have common features.

Striking variables often relate directly to time. For example, the gestation period for an elephant in 640 days, whereas for a mouse it is 20 days. Also, the time to reach reproductive maturity varies enormously, as does the rate and number of offspring produced in a beneficial environment. As we all know, temporal differences extend to maximum longevities. For a mouse, this is about 3 years; for man it is close to a century. It is often said that one year in a dog ís life is roughly equivalent to seven in a man or woman's life, but that does not apply to development to the adult, which in relative terms, is far more rapid in the dog. Also, from the keeping of pets, it is well known that the signs of ageing in a dog or cat, are not obviously dissimilar to those in a human. Old animals move more slowly, lose muscular strength, skin changes may affect coat colour or density, cataracts can develop and hearing may diminish. The internal changes are comparable too, and a good example is the loss of elasticity of the protein collagen. This is the commonest protein in the body, and as animals age it becomes progressively cross-linked. The rate of cross-linking in a mouse or rat is enormously faster than it is in humans. The many changes in tissues and organs with age are also associated with the onset of various pathologies, at least in developed countries. For dogs and cats, this is demonstrated by the increasing veterinary expenses, and for man the sharply increasing costs of health care as we get older.

The important facts are, first, that the rate of ageing varies enormously amongst mammals, and second, that the major manifestations of ageing, at the anatomical and physiological level, are remarkably similar. How can this be explained? In the previous chapter, we saw that the energy resources available to an animal such as a mammal, are broadly partitioned into three areas. First, the general metabolism which is common to all mammalian species. Second, all functions associated with development to adulthood and reproduction, and third, efficient maintenance of the adult body for a considerable proportion of the total lifespan. We know that patterns of reproduction vary greatly amongst mammals, but more specifically than that, we can see that the rates of reproduction, and the potential number of offspring that can be produced also vary enormously.

Before developing this theme further, it is very important to make the distinction between an animal's life in a natural environment, and a life in a protected environment, such as a zoo, or under domestication. As I have repeatedly stressed, animals live in hazardous natural environments. Some offspring do not survive long after birth; others never reach adulthood; some that do so may not breed, and those that do breed are unlikely to have nearly as many offspring as they are potentially capable of producing. In contrast, animals living in a protected environment, which provides food *ad libitum* and the absence of predators, have very different reproductive success, provided of course, that they have the freedom to breed. Infant mortality is usually quite low, most offspring develop to full adulthood and will continue to breed, unless human and other intervention prevents them from doing so. Under these conditions the important reproductive parameters are 1) the time it takes to reach reproductive age, 2) the length of gestation 3) the litter size and 4) the intervals between litters. To this must be added the period of fertility, which obviously relates to maximum lifespan. It is evident that mammals such as mice, rats and many other rodents, as well as some larger mammals such as rabbits, have a very high reproductive potential in the range 70-150 offspring per lifetime. At the other end of the scale, mammals such as the elephant, rhinoceros, hippopotamus, whales, the great apes and women are capable of producing relatively few offspring per lifetime, roughly in the range 8-12. (This takes into account the common effect of lactation in reducing fertility). It becomes obvious that in an unprotected natural environment, slow breeding animals must survive longer than fast breeding ones, if the population is to stay approximately constant in size. For a female elephant to produce two offspring which reach adulthood, she will need to give birth to at least three (taking into account infant and child mortality), and have a lifespan of nearly 30 years, which is about 40% of the maximum lifespan. On the other hand, a mouse need survive only about 12 weeks (assuming that on average only one litter is produced), which is less than 10% of the maximum lifespan of 3 years. A survey of a wide range of mammalian species demonstrates beyond question that there is an inverse relationship between maximum reproductive potential and maximum lifespan. This is illustrated in Figure 5.

Many studies have been published which attempt to relate maximum lifespan to the rate of metabolism of each mammalian species, or the size of the animal, or the size of its brain, or some combination of these values. (Others have calculated the total number of heart-beats in a maximum lifespan, and have even claimed this is broadly constant). Although some such relationships have been demonstrated, there are also some glaring exceptions. In particular, bats have evolved a unique lifestyle, in which they roost in inaccessible places during the day, and use their flying ability to forage widely for food at night. They can therefore evade predators and survive for longer in a natural environment than ground living mammals of comparable size. Yet bats have a metabolic rate comparable to other mammals of the same size. What is striking is their slow rate of reproduction and also their long lifespans. Many species produce only one offspring per year, and their maximum lifespan can be more than 20 years. Thus, bats are much more like larger mammals with regard to reproductive rate and longevity. Not surprisingly, in these respects they are comparable to many birds, which breed much more slowly than ground-living mammals of the same size, and have very significantly longer lifespans.

What we see amongst mammals is an uneven allocation of resources to reproduction. Given a constant proportion devoted to normal metabolic activities, this means that some mammalian species devote more resources to maintaining the body during adulthood than do others. It follows that maintenance mechanisms should be more efficient in long-lived species than in short-lived ones. This has indeed been very clearly demonstrated by a variety of different means.

Normal cells grown in the laboratory (so-called "tissue culture") have a finite lifespan. This was first shown for human cells, known as fibroblasts, which are an important component of skin and connective

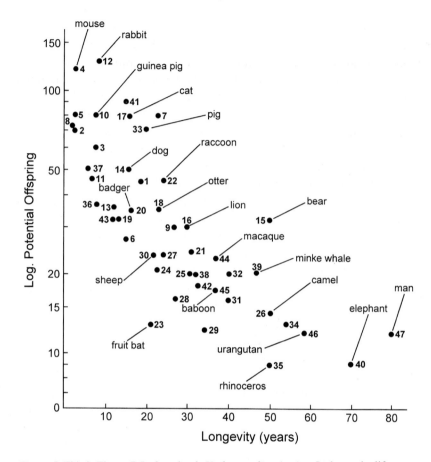

*Figure 5.* This is Figure 7.5 of my book *Understanding Ageing*. It shows the lifespanm of 47 mammalian species plotted against maximum reproductive potential under ideal conditions. Only 21 representative species are labelled. Note that the north American brown bear has a longer than expected longevity because in hibernates several months each year. The fruit bat has a lifestyle more like a bird than a ground-living mammal, and a lower than expected reproductive rate

tissues. These cells divide about 60 times, before they become senescent and cease further growth. It turns out that this is the longest lifespan of fibroblasts of all mammalian species subsequently examined. In fact, there is a clear relationship between fibroblast lifespan, expressed in cell divisions, and the maximum lifespan of the species from which the same type of cell were obtained. Cell maintenance must relate to the ability of cells to grow in culture (often referred to as their growth potential), whatever the actual causes of senescence may be. There is another very important observation on cultured cells. Mouse or rat fibroblasts grow for about 10 cell divisions before being senescent. However, these cells commonly change into cancer cells which can grow indefinitely. From a laboratory population of senescent normal cells, individual colonies or clones of cells with altered morphology appear. These cells have a variable number of chromosomes, unlike the constant number in normal cells, and they have become immortal, just like the cancer cells that can be grown from tumours. The important point is that normal rodent cells are not well protected against their conversion to abnormal types of cell. In contrast, normal human cells when they become senescent in culture have never been known to produce immortal derivatives. They are completely stable, so are obviously endowed with one or more mechanisms to prevent this happening.

This is indeed just what one might expect from the study of the whole animal. Three year old mice often develop tumours which may kill them, or, if they die from another cause at this age, may still have incipient malignant tumours in this or that tissue. The same is true of 70-80 year old people; they may die of cancer, or they may have incipient tumours. The probability of a normal human cell developing into a tumour is enormously lower than it is in a mouse. Or to put it another way, if human cells were like mouse cells, we would certainly all die of cancer in infancy or early during childhood. Clearly those mechanisms which maintain the normal properties of cells in the body, are far more effective in man than in the mouse. The incidence of carcinomas in the rat (with a lifespan of about three years) and humans is illustrated in Figure 6.

The reasons why cancer cells become immortalised has been the subject of intense research for many years. One property they have acquired is the ability to maintain the normal ends of chromosomes, known as telomeres. Most normal body cells do not have this ability, so their chromosome become very slightly shorter everytime the cells divide. This is thought to be one cause of the senescence of dividing cells, but it is not necessarily the only one. Germ-line cells are potentially immortal, and it is known they maintain their telomeres. It is also likely that they have other important maintenance mechanisms, lacking in somatic cells.

A number of other maintenance features of mammalian body cells have also been studied in detail. One is the ability of cells to repair damage in DNA, induced by ultra-violet light. Several studies have

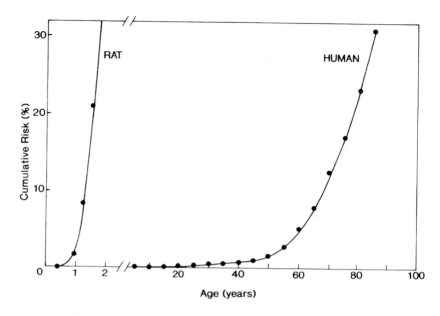

*Figure 6.* The age of onset of cancers (carcinomas) in the rat and in humans

confirmed that normal cells from a range of species with different maximum longevities, have a range of repair efficiencies, that it, there is a direct correlation between repair capability and longevity. One study is of particular interest, because the cells were taken from the outer layer of cells of the lens of the eye. These cells are, of course, directly or indirectly exposed to sunlight, so they have a specific need for this repair mechanism. It turns out that the cells which have the longest exposure to sunlight ie. through the animals' lifespan, also have the most efficient repair. For six species, with maximum lifespans ranging from 3 to 40 years, the correlation was excellent. The levels of enzymes involved in other aspects of DNA metabolism, have also been shown to be correlated with maximum longevity. There is also evidence that damage to DNA from oxygen free radicals is greater in shorter lived species, which probably means that the defences against free radicals are more efficient in long lived ones. This is confirmed by a study of the amount of the carotenoid anti-oxidant in blood serum, which correlates with longevity, and also other aspects of free radical elimination. Another maintenance mechanism depends on the oxidation of toxic chemicals. Again, evidence has been obtained that longer lived species have more efficient detoxification mechanisms than shorter lived ones. An interesting study compared the long-lived pigeon and the short-lived rat, which have similar metabolic rates. The former was much more efficient in defending itself against free radical damage than the latter. Similarly the long lived paraqueet and canary suffer less damage than the mouse.

The Australian scientist Alec Morley was the first to measure the increase in somatic mutations during human ageing. These mutations were detected in blood lymphocytes. When he did similar experiments in mice, he found approximately the same increase during their lifespan. However, mice live for only about 3 years and humans 30 time longer. Therefore the mutation *rate* is about thirty fold higher in mouse than in man. This is complelety in line with other comparative studies.

Many of the published studies I have just summarised, and several others I did not mention, depend on sophisticated techniques, which would be difficult to explain or outline here; nor would it be appropriate to list nearly 20 different investigations. The important conclusion is that in almost every case the expected relationship between efficiency of maintenance and longevity has been demonstrated. The techniques are now available to carry out even more investigations using animals with different lifespans. The strong prediction is that the correlation between the efficiency of cell or tissue maintenance and maximum longevity would receive even stronger confirmation.

In some instances the origin of age-related changes in molecules such as proteins is not well understood. A good example is the progressive cross-linking of collagen which occurs during our lifespan. What is clear is that the cross-linking of tendon collagen in the rat over a three year period is far faster than it is in human collagen. At first sight this seems surprising, because rat and human collagen are chemically extremely similar, and have exactly the same functions in the body. What we must conclude is that the milieu in which these protein molecules exist is different. Proteins can be modified in many ways by particular chemical reactions, involving oxidation, addition of sugars, changes in particular amino acids, and so on. These reactions are subject to changes in rate, and it is not unreasonable to suppose that the evolution of greater longevity, depended on an improvement in the control of unwanted side-reactions which damage proteins. For this reason, proteins in many locations survive undamaged for much longer periods in long-lived animals than in short-lived ones, even though their intrinsic structure and properties are the same. The crystallins of the eye lens loses transparency through side reactions, which eventually cause cataracts to form. Most old animals are liable to develop cataracts, but the rate this happens is far slower in man than it is in dogs, and slower in dogs than it is in rats or mice. Similar conclusions can be drawn about the progressive damaging changes in the vascular systems of mammals with very different longevities.

All the evidence is compatible with the general view that cell and tissue maintenance is essential for the normal survival of the adult organism, and that ageing is associated with the eventual failure of such maintenance. Moreover, the resources invested in maintenance are inversely related to those invested in reproduction. There is a trade-off between the two: as one increases, the other decreases, and vice versa. The balance is set by the particular adaptation and life style of the species in question. Those animals which live in hostile dangerous environments must breed fast to maintain their numbers. In consequence they invest less in maintenance and have short lifespans. Those animals which are successful at evading predators and finding adequate supplies of food, have a much lower annual mortality, can afford to reduce fecundity and still maintain the population size. However, to achieve this end, they also have to stretch out their lifespans and this depends on the evolution of better maintenance mechanisms. All these changes occur very slowly under the influence of natural selection, and I will explore this issue further later on.

There is an established fact about the ageing of mice and rats, which is very pertinent to the general theme I have developed. When such animals are given a restricted diet, normally about 60% of what they would eat if fed *ad libitum* (that is, being free to eat as much as they want), their lifespan is increased very significantly. In many different studies, a life-extension of about 50% is consistently observed. Moreover, these animals have significantly fewer age-associated pathologies than animals given a full diet. What is not always realised is that these calorie-deprived animals also become infertile, or have very low fertility. This makes biological sense, because small ground living rodents do not have a constant supply of food in their natural environment. Sometimes there is a glut, and at other times the animals are liable to become semi-starved. In the absence of nutritional resources, it would be pointless to attempt to breed, since this would endanger both the mothers and her offsprings' chances of survival. So the first thing that happens when food supply runs out, is the shutting down of reproduction. Clearly this is an adaptive mechanism. When the food supply is restored, breeding begins again. In fact, it has been established experimentally that when calorie-deprived animals are subsequently allowed to have a full diet they can breed at very significantly greater ages than animals which are continuously fed *ad libitum*. It also makes biological sense that the semi-starved animals invest what resources they have (namely, about 60% of the normal calorie intake) in maintaining their bodies, until such time as more food becomes available and breeding can begin again. The diversion of resources from breeding to mainte-nance, has the effect stretching out survival, that is, increasing the lifespan. Seen in this light, the extension of lifespan is an evolutionary adaptation, which maximises the fitness of the animal (i.e. the total number of offspring produced) in environments with fluctuating food supplies. One can look at the situation in another way. One might guess that a rat or mouse given 60% of a normal diet, would simply keep breeding but produce fewer offspring. But this does not happen; instead, the animal has the positive response to reduced diet of not breeding at all.

It has often been asked whether calorie deprivation would also increase the longevity of people. The short answer is that we do not know. We do know that excessive use of available energy, for example by ballet dancers, gymnasts, and athletes, can result in a cessation of menstruation in women. But we do not know whether a lifetime, or part of a lifetime, with a significantly low intake of calories increases longevity. One might think this information would emerge from the examination of populations of people with different nutri-tional lifestyles. The problem is that the many human populations with inadequate food in third world countries also have inadequate health care, and therefore a lower expectation of life. Similarly good health care is almost always associated with full availability of food. Never-theless, I think it quite possible that a long-term study of individuals with low and high calorie diets within one such community would indicate whether man responds in the same way as mice and rats. What is already clear is that a sustained low-calorie diet is likely to have detrimental side effects, and those who have referred to the possible "beneficial" effects of calorie-restriction seem to have forgotten that it may well reduce fertility, and probably libido as well.

# Chapter 7. How Many Genes?

Each human unfertilised egg contains one complete human DNA genome in its 23 chromosomes, and each human sperm cell contains a human genome, with an equal chance it will have an X sex chromosome or a Y sex chromosome. On fertilisation there are two genomes, one from each parent, but the females have two X chromsomes and the male has an X and a Y. When the embryo develops, each cell has these two DNA genomes, and this is true of every cell in our body, except for red blood cells. Some cells of the immune system have a slightly altered genome. Also, there is a small amount of DNA in the respiratory mirochondria, which are inherited from the mother.

Each genome has about 3 billion ($3 \times 10^9$) bits of information, that is, the basic units of information abbreviated to A, T, G & C (see Glossary). The human genome project was the unravelling of the complete sequence of these units. Many people believe we will then know how many different genes we have, but I suspect this will only be partly true, because to count the genes we have a definition of what a gene is, and there may be more types of genes than are at present suspected. What most scientists are usually referring to when they say we will know the number of genes, are the genes that define the structures of all the proteins in the body. It is now known that there about 30,000 of these per genome, somewhat fewer than was originally suspected Although large, such a number only accounts for a fairly small proportion of the total genome. Many scientists believe that much of the rest is just "junk" DNA; others, including myself, think that at least a proportion of this DNA will turn out to have other functions, at present unknown, or poorly understood.

Two major facts are non-controversial, first, that the DNA contains all the instructions for making a human being, and second, genes specify the structure of all the different proteins in our body. Since each mammalian species has a maximum lifespan, preceded by a period of senescence, there must be genes that in some way control ageing and longevity. Some believe that ageing is genetically programmed, or even more strongly, that there must be a programme for ageing. This view is far too simplistic, and I have cited the ageing of teeth as a simple illustration. The size and structure of teeth is genetically determined, but teeth wear down or decay with use. Teeth are indeed programmed to last a lifetime, but if, hypothetically, they were not used they could last much longer than a lifetime. The whole architecture of the body is determined by genes, but this design, as we have seen, is fundamentally incompatible with indefinite survival. Also, the major maintenance mechanisms are all specified by genes. For example, the efficiency of DNA repair depends on the success of enzymes, whose structure is determined by genes, and there are at least 150 repair enzymes. For maintenance mechanisms *in toto*, several thousands of genes are involved. Ask any immunologist how many genes specify the whole of our immune system, and the number suggested will be a large one, probably more than a thousand Similarly, the numbers of detoxifying enzymes specified by genes is already known to be very large. All these genes, in one way or another, contribute to the determination of lifespan.

This is not to say that there are not important random events which damage molecules and cells and contribute to ageing. There may be rare types of damage in DNA which are not even recognised by repair enzymes, or there could be critical errors in molecules which have lasting effects. It is possible that certain deleterious errors in proteins precipitate further changes of the same type, which leads to progressively damaging effects, for example, in brain cells. We know that there is a strong random or chance element in the development of malignant cancer. One individual can be unlucky, and develop the disease unusually early, whereas others remain completely free of any tumour. Exposure to an infection early in life may trigger a strong immunity, which serves that individual well throughout life. Other individuals, without such protection, may succumb to the same infection when their immune system declines with age.

There are good reasons to believe that random events are an important contribution to ageing. When animals are inbred, the small differences between maternal and paternal sets of genes are ironed out. Normally this is harmful, as hidden or recessive deleterious genes can become expressed. However, scientists have been very successful in inbreeding several strains of mice. These mice have become genetically identical, as similar to each other, in fact, as identical twins in man. Groups of these animals have been used to study their lifespan. Each strain has a characteristic average lifespan as we might expect. What is remarkable, however, is that the lifespan of the animals of any one strain is quite variable (for example, from about 21 to 34 months for male animals of the fairly long-lived inbred CBA strain, leaving aside a few early deaths). It is remarkable because we know all the genes are the same, and the animals are kept in a very uniform environment with a constant and defined supply of food. By a process of elimination, we must conclude that intrinsic chance events are very important in determining lifespan. The same conclusion comes from the study of identical twins. Although their lifespans are more similar than normal sibs, these are commonly substantial differences, which in this case could, of course, be attributed to environmental effects.

In spite of the major conclusion that very large numbers of genes influence in one way or another an animals lifespan, it is commonly believed that there are a few critical "gerontogenes" which really determine lifespan. One reason is that gene mutations have been identified, for example, in a particular small roundworm or nematode, which considerably increase lifespan. Those who have done such experiments believe they have identified single genes that in some way control lifespan. One should be very cautious about such a conclusion, because there are several ways one or a few genes could increase longevity, which I mention below. Another reason that some believe there may be only a small number of genes that control ageing, come from experiments with the fruit fly Drosophila, an organism famous for revealing the secrets of heredity. It has been possible to increase the lifespan, simply by repeatedly selecting the eggs laid by the eldest mothers in a population. This experiment depends on the fact that the flies have two copies of every gene, and these copies are often slightly different from each other. In selecting eggs from the eldest mothers, the experimentalist is exploiting variability in these genes. He is in effect, seeking out those genes (or more strictly, the best of each pair of genes) which have an effect on lifespan. The ultimate result, after about 15 generations of selection, is a significant increase in the lifespan of the whole population. The conclusion is that there are individual genes that have an effect on lifespan, and possibly that the number of such genes is not large. All this is correct, but we need to look more carefully at the question of genes and ageing.

There are in man known heritable traits which result in premature ageing, but luckily they are very rare. One is known as progeria. Young children with this genetic defect are initially normal, but after a few years a series of abnormal features become apparent. These include short stature, loss of hair, skin changes which make the face look adult-like, and several abnormalities in the vascular system which normally cause death in the early teens. Another less severe defect is called Werner's syndrome. Again, the individuals initially appear normal, at least until the second or third decade of life. But then abnormalities set is, including whitening of the hair, progressive skin changes, poor wound-healing, cataracts, diabetes, increased risk of cancer and heart disease. The average lifespan is 46 years. It has been known for a long time, that Werner's syndrome is due to a single recessive inherited defect, that is, the defective gene must be inherited from both parents. Many scientists working on this gene believed that if its function could be identified, a great deal would be learned about the process of ageing. Some even believed it might be a gerontogene which would be the key to understanding ageing. The gene has now been isolated and shown to be like one determining the structure of a protein important for unwinding the double helix of DNA, known as a helicase. How can a single defective gene like this have multiple and diverse effects on the individual? One simple explanation is to think of the defective gene product as a "spanner in the works". Any complex machine can grind to a halt if there is a single defective component. Or there may be a single essential component which does not last as long as it should, so the whole machine's lifespan is reduced. With regard to Werner's syndrome, the likelihood is that the defective helicase is one of several helicases. Initially these are able to fulfill all normal functions, but as time goes on DNA repair and maintenance is not quite all it should be, so a wide range of body features become progressively affected. So, unfortunately, knowledge of the defective gene causing Werner's syndrome has not provided as much insight into the cause of ageing as might have been hoped, because there are multiple causes of ageing and multiple genes involved.

Before the human genome was sequenced, a large number of human genes had already been identified from the study of human genetics. Human populations are very large and our medical screening programmes are very efficient. A very wide variety of medical defects were known to be inherited from the study of family pedigrees, and a high proportion of these defects are due to mutations in single genes. In many cases, the defect at the biochemical level has been identified. Some years ago, the pathologist and gerontologist, George Martin, went through the whole catalogue of human genetic defects, and listed all those that had some connection with ageing or age-associated disease. The proportion of these genes out of the total number known at that time was surprisingly large, and now more and more are being identified, many being necessary for the maintenance mechanisms discussed in Chapter 3. Thus, the conclusion is that many genes, in one way or another, have something to do with the changes that occur during ageing. This supports the view that ageing is multicausal or multifactorial, and is contrary to the notion that there are just a few "critical genes" which determine lifespan.

What about the genes in worms or flies which increase longevity? At first sight, they seem to play a critical role in the ageing process, because changing one gene can increase lifespan by as much as 50%. This is very much like taking a complex machine, changing one component, and then finding the machine lasts twice as long as before. With regard to organisms, all sorts of possibilities exist. For instance, we know that reducing calorie intake extends lifespan in rodents. It is fairly easy to imagine a mutation which simply reduces the efficiency of digestion, thereby decreasing the intake of calories, and thereby having the long term effect of increasing lifespan. Alternatively, we can easily envisage genes which reduce fertility, and allow more resources to be diverted to maintenance. There could be genes that slow down metabolic rate, reduce the production of ROS, and as a consequence there is less damage inflicted on proteins and DNA. All these possible changes, and probably many others, are important and are relevant to ageing. However, from a biological point of view, they are in a sense experimental artifacts, because one could predict that the mutants produced in a laboratory environment, would not be advantageous in a natural one. In all cases, the mutants would probably be selected against and be eliminated because they were less fertile, less active or mobile, or whatever. Of course, much the same could be said about the artificial selection of animals. Many domestic animals, selected for particular desirable traits, would do badly in a natural environment in competition with the "wild type" animal from which it was derived. In fact, it is known that domesticated animals introduced into a natural environment (such as the pigs introduced by Captain Cook into New Zealand) soon change by natural selection back to the original form of the species. Feral cats in Australia have often become much larger than the domestic variety.

The genes identified by scientists which affect the longevity of experimental animals probably have less to do with the genetic control of ageing than they would like to think. Nevertheless, during evolution genes must have been selected which affect longevity, because we know in different instances that the longevity of certain animals decreased during their evolution, whereas that of others increased (as we shall see in the next chapter). There are two ways these changes could occur. First, we know that there is intrinsic variability in any animal population, affecting a wide variety of animal features. We know that genes inherited from the maternal parent are often slightly different form that those inherited from the paternal parent (They have slightly different DNA sequences). Artificial selection for almost any feature of the animal acts on this variability and over a few generations significant changes are seen. For example, selecting for larger or smaller size in dogs will produce very significant changes in weight and dimensions. The experiment previously mentioned, in which the eldest egg-laying female fruit flies were repeatedly selected, depends on the intrinsic variability in genes. The selection acts on new combinations of genes, to produce in a few generations flies with significantly increased lifespan. Human populations have a similar intrinsic variability affecting lifespan. It is well known that some families tend to be long-lived and others shorter lived. It is joked that if one wants to have a long lifespan, one should choose long-lived parents and grandparents. There are a few individuals who have lived to 115–120 years of age, and it is probable that these, by chance, have combinations of genes which delay senescence and ageing. They exist in human populations, because such populations are extremely large, so only a minute proportion of individuals have these combinations of genes. (I refer here to populations where the birth date of the very elderly individual is properly documented. There are of course innumerable claims of extreme age, where there is no authentic documentation of birth). The second type of variability which can affect a given feature of an animal depends on new mutations in genes. Thus, the huge differences in size between different breeds of dogs, from the Great Dane to the Chiwawa, is due both to new gene mutations, as well as intrinsic variability, both of which are selected for artificially. We can confidently expect that there are mutations which will increase lifespan, as well as those that reduce it. The ways these new mutations operate is clearly a very complex problem, especially as I have stressed that ageing has multiple causes. A mutation which reduced the rate of collagen cross-linking would have little or no effect on the ageing of the brain, the eye and so on. However, evolution takes place over a very long period of time and it is not impossible to envisage genetic changes, each of which have a very small effect, adding up to a very significant increase, or decrease, in the overall rate of ageing. Moreover, there could be mutations which have multiple effects on cells and tissues of the body. For example, a mutation which reduced the production of ROS might have multiple effects. An improved defence against ROS would have a similar outcome. We can easily envisage changes in the efficiency of DNA repair. Suppose, for example, a gene producing a DNA repair enzyme is duplicated. This could be followed by the evolutionary divergence of the two genes to produce repair enzymes with related but overlapping functions. We know that DNA repair is more efficient in human cells than in mouse cells. This could be due to the greater efficiency of a particular pathway of repair, or more likely, it could be due to more alternative pathways. The duplication of particular functions is well known in evolution; if one fails for any reason, there is very commonly a back-up mechanism. With regard to the many maintenance mechanisms previously discussed, the prediction would be that long-lived species have many more back-up mechanisms than short-lived ones.

Another type of mutation may also be very important in evolution. We know that there is some relationship between the rate of development of an animal to its maximum lifespan. In general, long-lived species become sexually mature more slowly than short-lived ones. Therefore, if a mutation acts to slow down the rate of development it is likely to also increase the lifespan as well. The mutation acts to stretch out the whole life cycle between generations, including the period of gestation, growth from infant to adult, period of active reproduction, and the final stage of senescence and death. The domestic cat and lion are quite closely related, but all these stages happen much more quickly in the domestic cat than the lion. For cats and lions the actual values are respectively: gestation period, 65 days and about 110 days; time to reproduction, 10–12 months and 3–4 years; inter litter interval 6 months and 1.5–2 years; longevity 15 years and 30 years. These are all quantitative changes which could certainly be caused by a fairly small number of mutations. It is even said that the mew of a cat when recorded and slowed down, sounds much like the roar of a lion, which is again another fairly simple quantitative change.

There is a special evolutionary mechanism, known in several zoological contexts, where the adult of an evolved species resembles the not-yet adult of form of an earlier species. It is known as neoteny, and it depends on the later species becoming sexually precocious in a body which, anatomically speaking, is not yet fully mature. A strong case can be made for the occurrence of neoteny in the evolution of man, because anatomically our adult bodies resemble more strongly the infants of great apes than adult forms. Mutations which have quantitative effects on development of the body and the development of sexual maturity would be responsible for neoteny, and it does not need much imagination to envisage an effect on lifespan as well. The novelist Aldous Huxley exploited this theme in his novel *After Many a Summer*. A particular diet induced great longevity in two individuals, but unfortunately these individuals also lost their normal human characteristics and developed ape-like features when they became very old.

There is an argument put forward by the gerontologist Richard Cutler that longevity evolved very rapidly during the evolution of man from higher primates. Curiously, Cutler took the maximum longevity of human to be 120 years, and that of the great apes as 60 years. Now it is true that gorillas, chimpanzees and orangutans live for up to 60 years in the well protected environment of a zoo. However, the number of animals which have actually been kept from early age for such a long period in any zoo is quite small, probably less than a dozen. In contrast, the number of recorded ages of humans in countries with proper records is vast. Amongst these are a few who have lived for 120 years. Now, imagine a dozen human individuals selected at random and kept in zoo-like conditions. How long would they live? My guess is that it would be close to the average human expectation of life in Westernised societies, say 70–80 years. So the increase in lifespan during human evolution is not a doubling, as Cutler assumes, but an increase of about 30%. Such a change occurred over a period of about five million years (the time the chimpanzee and human lineages diverged), perhaps around 200,000 generations. This is quite a long time for natural selection to act, first on intrinsic genetic variability, second on new mutations affecting maintenance, or having quantitative effects on the human life-cycle. The reasons why human beings have such a long lifespan is an interesting and important one which will be outlined in the following chapter.

# Chapter 8. The Evolution of Human Longevity

From a simple biological standpoint, it is remarkable that *Homo sapiens* is the longest lived terrestrial mammalian species. The largest, such as the elephant, hippopotomas and rhinoceros, have long lifespans, but not as long as human beings. There are herbivores which are much larger than man, and also carnivores, but their lifespans are well under half the human one. Scientists studying ageing have correlated brain size, or the ratio of brain weight to body weight, with lifespan. There is in fact a good reason why, at least for humans, there should be an association of brain size with longevity, and this will be explained in this chapter. Another feature of human ageing is the well known fact that women have a longer expectation of life than men, by 5–7 years in developed countries. This has so far never been satisfactorily explained, but I suggest an explanation here.

There are two major evolutionary features of ageing. The first is the reason for the evolution of ageing *per se*, which I discussed in Chapter 5. The second is the determination of the actual lifespan, or the evolution of longevity of any given species. Starting with a given lifespan, what are the contrasting evolutionary forces which result either in an increase or decrease in that lifespan? To fully understand how natural selection can alter longevity, it is easier to first consider the case in which maximum lifespan gradually decreases during evolution.

In natural environments, it is common for the number of individuals in a population to remain roughly constant over quite long periods of time. This steady state in animal numbers is due on the one hand to natural attrition, or the annual mortality, and on the other to the rate of reproduction. For a population to survive, each pair of breeding adults must produce, on average, another pair of breeding adults. Now imagine an increase in the annual mortality, perhaps due to the appearance of a new and more effective predator, a shortage of food, or to some other change in the environment. This means that the survival of offspring, as well as parents, is reduced. The number of animals in the population will gradually decline, and if this trend continues the population in question will become extinct. As we saw in the last Chapter, genetic variation exists in all populations, and this variation affects reproductive parameters, as well as survival parameters. In a declining population, the animals which will be favoured by natural selection are those that produce more offspring, or offspring at a younger age. To do this, more resources must be invested in reproduction, which means, as we have seen, less can be invested in maintenance. Thus, the favoured animals will be those that develop faster, reproduce more quickly, but also have a slightly shorter lifespan. Over a long period of time, the average longevity of the population or species becomes shorter.

A specific example can be cited from a major group of mammals, the carnivores. The smallest carnivores, such as stoats and weasels, have a very active lifestyle, because they depend entirely on a continual supply of prey. In reproducing they have a short gestation period and large litters. The offspring develop rapidly and become sexually mature at an early age. These animals have the shortest lifespans of any of the carnivores, only about 5–6 years in captivity. It could also be added that their annual mortality in a natural environment is very high, the main cause of mortality being a failure to find prey. One might say they lead a precarious existence, but they also have remarkable adaptations to their particular ecological niche.

One of the best documented evolutionary changes is that which occurred during the emergence of the modern horse. Its ancestor was about the size of a hare, but over a long period of evolutionary time its size gradually increased. We can also assume that its rate of development decreased, and probably its rate of reproduction. Although it can never be measured, it is also likely that its longevity increased.

The same trend can be seen amongst many species of primates. Small monkeys such as the marmosets, the squirrel monkey, as well as small lemurs and loris species, begin to reproduce at the age of 1–2 years, and some of them have 2–3 offspring per litter. Their lifespans in captivity are in the range 10–15 years. Larger monkeys such as the macaque, spider monkey and baboon develop more slowly and begin reproduction at the age of 5–6 years, producing only one offspring per pregnancy. Their lifespans in zoos is in the range 30–40 years. Moving up the evolutionary scale, we find that the great apes (gorilla, chimpanzee and orangutan) develop even more slowly, reaching reproductive maturity at 8–14 years, with a lifespan of 50–60 years. The trend continues to human beings with an average age of reproductive maturity of about 16 years, and of a expectation of life of 80 years or more. The higher rates of reproduction of the smaller primates correlates with a greater animal mortality. The gradual evolution of the primate is associated with more successful adaptation to the environment, lower rates of reproduction and longer lifespan.

What were the evolutionary forces which resulted in the emergence of long-lived hominids from ape-like species? The key is successful adaptation to the environment. As adaptation improves, mortality falls and the result is that individuals in the community tend to survive longer. Both the great apes and humans reproduce quite slowly. The interbirth interval is often 2–3 years, but this depends on infant survival. The infant mortality rate is about 25% in natural populations in the great apes, and in the absence of lactation females become pregnant quite quickly. However, the successful suckling of an infant suppresses fertility, so the interbirth interval is extended. In humans, reproduction usually starts at 16–18 years, depending in part on the health and nutrition of the individual. The average interbirth interval is about 2 years, and a female that survived to 28 years, would produce, again on average, about 6 offspring. With these reproduction parameters it is possible to calculate that the population size would be maintained with an annual mortality of about 7% per year after infancy. It can also be calculated that the expectation of life at birth, assuming 25% of death in the first year and 7% thereafter, would be only about 18 years. The expectation of life of females who reached the age of reproduction would be about 28 years. The number of individuals who reached the age of 45 would be about 3% of the population. Obviously, such hominid populations lived in harsh conditions, where the high annual mortality would largely be due to disease, starvation and predators. One could question the accuracy of the estimates of annual mortality and overall survival, but I believe that with known reproductive parameters, they are in the right range. It could be argued that the development of adult experience and skills, would tend to reduce annual mortality. This might have a small demographic effect, but will not alter the overall picture of human evolution. (Even in the advanced community of Ancient Greece, the study of skeletons in burial grounds shows that survival after infancy is close to exponential, that is, constant annual mortality over the lifespan). It should also be noted that infanticide, if practiced, will have a fairly small demographic effect, because women will subsequently become pregnant more quickly.

Under the circumstances outlined, how does natural selection act on these early hominid populations? Clearly the females which survive the longest produce the most offspring, so any gene that favours survival, will be transmitted more commonly than genes that have the contrary effect. That is simple Darwinism. In the case of hominids, survival depends on successful adaptation to the environment. The hunter-gatherer life style combines a vegetarian diet with the eating of meat. In hunting, success depends on co-operation between members of the social group, which could be one extended family, or a few such families. Co-operation depends on communication, so it is obvious that language is an advantage. Also, transmission of information and skills from one generation to the next increases the likely survival of each new generation. This would include the first use of tools, and then their further development and elaboration. Also, it has been suggested that the departure from an arboreal forest habitat to more open ground-based one, would allow women with young offspring to have greater use of their arms and hands, during, for example, food gathering. There would also be greater use of arms and hands with the evolution of an upright, bipedal stance and mobility. An omnivorous diet, instead of a vegetarian one, does not require the large abdomen seen in great apes. This in turn allows humans to run, and thereby capture prey more successfully, as well as diminishing the success of predators All these changes may well have occurred in concert, and would also have been accompanied by a gradual increase in brain size.

There are several major consequences of an improvement in life style and adaptation. First, the annual mortality would become less. Second, more individuals would survive to reproductive age. Third, amongst reproducing females, those that survived longest would on average produce the most offspring. There would therefore be selection for genes which favour late survival and late reproduction. These are the type of genes which were mentioned in the previous chapter, namely, genes which favour the whole stretching out of the life cycle, especially later reproduction and longevity. This is exactly similar to the experiment with fruit flies: when the last egg laying females were repeatedly selected, longevity of the population increased. In this laboratory experiment a significant increase has seen over 15 generations: in the case of hominid evolution, there would be thousands of generations during which natural selection would act. Another feature of hominid evolution would also have been important. The development of language and the transmission of information and skills from one generation to the next, would favour young individuals who could absorb all these advantages. The period of learning during development to adulthood became much more important than it is in pre-hominid animals, so slower growth and development to reproductive maturity would be favourable traits. Although this would tend to reduce overall reproductive potential, it would be compensated for by the greater reproductive success of older individuals. Taken together, all these evolutionary forces would act to increase maximum lifespan, and there is therefore a real reason why large brain size is associated with longevity. To put it another way, it is not fortuitous that the cleverest and most successful mammalian terrestrial species is also the longest lived.

It might be thought that reduced annual mortality would result in a dramatic increase in population size, but this is not necessarily the case. There would in fact be two opposing trends in these early hominid populations. The reduction in annual mortality would indeed tend to increase survival and an increase in population size, but at the same time the reproduction of younger individuals is being decreased, so the fecundity of the population as a whole decreases. The balance between these two trends will depend on other ecological factors, such as the food supply. This can be illustrated by the ancient aboriginal communities which lived in Australia for at least 40,000 years. It is estimated that the number of individuals in 1788, when colonization began, was in the range 500,000 – 1 million for the whole continent. Suppose the founder population arriving from the North was 1000 individuals, and the average generation time is 20–25 years. It can be calculated that over the 2000 or so generations that elapsed until 1788, the increase in population size per generation is only about 0.25%. In effect, the population is in a steady state. This situation is almost certainly due to the harsh conditions in which these communities lived. The limited availability of food almost certainly was the main factor in setting a limit to the population size of individual tribes. In sharp contrast, the Polynesians had a plentiful supply of food, both from agriculture and efficient fishing, using sea-worthy canoes. In consequence, island populations rapidly increased, and as a result of population pressure in confined habitats, groups of individuals were forced to migrate by sea to find new habitats.

In the situation where an increase in longevity is evolving, selection for longevity is acting on females rather than on males. But almost all of our genes are present in both males and females, so an increase in the longevity of one sex will also almost equally affect the other. Nevertheless, there are some gene differences, as a direct consequence of a mechanism for sex-determination. Males have an X and Y sex chromosome, whereas females have two X chromosomes. There are believed to be only a few important male-determining genes on the Y chromosomes, and it has been said that the rest of the chromosome is genetically inactive. One cannot be sure about this, because it has recently been shown that DNA often produces RNA, which does not code for proteins, but has some other important function(s). Although in females only one of the two X chromosomes is fully active, there are short regions where genes in both X chromosmomes are active. Given that the selection of genes increasing longevity will occur to a greater extent in females than in males, it is therefore possible that there are a few genes on the Y chromosome which tend to reduce longevity, and also a few genes on the X chromosome, which in double copy tend to increase longevity. This would account for the greater expectation of life of females over males. Such differences, of course, could be over-ridden by other genes in the population which confer very long life-span on individuals irrespective of their sex. Some gerontologists are studying centenarians with the aim of identifying such genes. By the same token, sex-linked genes which affect lifespan might also be identified.

Although the age-related decline of reproduction in many mammalian species is sometimes referred to as menopausal, it is probably more appropriate to restrict the term menopause to describe the fairly sudden cessation of reproduction in human females. Clearly this is a programmed event, due to hormonal change, which is a much more distinct change than occurs in other mammalian species. Some have argued that the evolution of the menopause, occurring many years before final senescence and death, is hard to explain on evolutionary grounds because it has the effect of reducing reproductive potential. This view does not take account of a well established feature of evolution known as kin-selection, first recognised by the geneticist J.B.S. Haldane, and then developed much further by the evolutionary biologist W.D. Hamilton. Parents and children have 50% of their genes in common, as do siblings; and 25% of a grandchild's genes are present in each grandparent. The care of children by parents and grandparents (as well as by sibs or other relatives) increases the likelihood of the transmission of family genes. This is how kin-selection operates. In humans, children need extended parental care, so there would be an increasing burden on mothers, especially in a harsh environment, if they continued to produce offspring at later ages. It would be advantageous to cease reproduction at a certain time to allow existing offspring to be cared for. It would also be advantageous for non-reproducing women to spend time caring for their offspring's offspring, that is, their grandchildren. All this makes evolutionary sense, and it is therefore probable that the menopause evolved as an adaptation during the emergence of hunter-gatherer communities, based on altruistic behaviour amongst family groups. I suggested that in a steady state hominid or pre-hominid population, about 3% of individuals would be expected to reach the age of the menopause. In biological terms, this is a significant proportion. Moreover, as the human population increased in numbers and expectation of life also increased, a larger and larger proportion of women would expect to reach menopausal age, and these post-menopausal individuals would make an ever-increasing contribution to the survival of their own family members.

Earlier in this Chapter, I suggested that an environmental change that increased mortality would result in the selection of animals that develop and breed more quickly, and these would also have decreased longevity. Small animals that already have these properties could adapt rapidly to the changed environment. The situation is different for large slow breeding animals: when confronted with adverse conditions, they would adapt much less slowly and effectively. One could predict that many might subsequently become extinct. This is in fact is exactly what happened to the megafauna in Australia about 50,000 years ago. Similar extinctions of large mammalian species were seen in North America and elsewhere, but not in Africa. (Note that these events are independent of the way the environment changed for the worse. All that it required is an increased annual mortality).

From the argumants presented in this Chapter and elsewhere, many of the questions which have been raised by various authors over the years can now be answered. Why do human beings live longer than other land mammals? Why is there a connection between brain size, intelligence, and long lifespan? Why do women live longer than men? Why did the menopause evolve? The answers which have been provided depend entirely on the Darwinian interpretation of human evolution. My hope is that they will become common knowledge, and so make the questions unnecessary.

# Chapter 9. Myths of Life Extension

Ever since humans became fully aware that they will get old and die, individuals must have wondered about the possibility of extending longevity, or achieving immortality on earth. This has resulted in innumerable myths relating to supposed examples of life-extension, one of the most notable being the 969 year lifespan of Methuselah, as recorded in the Bible. Christian fundamentalists are liable to say that in those days people lived much longer than they do today! These are many other claims, and far too many to list here. The Guiness Book of Records has stated that maximum lifespans of people and animals are some of the hardest records to document. Some of the claims of human longevity have occurred in the 20th century, and as in some cases they were initially accepted by scientists, and they provide good examples of how people are deceived.

Many people living in the Caucasus region of the ex-Soviet Union claimed their ages were over 120 years, or even 160 years. There was even a touring troup of centenarian dancers, so it is not surprising the claims received wide publicity, and also wide acceptance. The oldest people were usually males which is intrinsically unlikely as human females live longer that males. Careful scrutiny of these cases, particularly by the gerontologist Zhores Medvedev, has shown that none are backed up by reliable records. In some cases, individuals took on the identity of their father, in others they exaggerated their age to escape military conscription, or they simply added on years, because the societies in which they lived respected the wisdom and experience of very old people. Another well-known set of claims of extreme old age was at Vilcabamba in South America. Again, some scientists (in this case anthropologists) believed the reports because the dates of baptism were recorded in the local church registers. More careful scrutiny showed that the entries usually referred to the individual's father, with exactly the same name, or even that of a grandfather.

The existence of properly authenticated birth certificates has made it possible to assess in the last century the true picture, at least in developed countries. There is no doubt that the number of centenarians is rising all the time, which is due to improved health care in these countries. However, the number of individuals reaching each additional year after 100 drops off rapidly, and it is quite rare for any centenarian to reach the second decade (ie. more than 110), and extremely few reach 115. The longest lifespan recorded with certainty is that of the French woman Jeanne Calment who died in 1997 at the age of 122 years. Although the improvement in medical care has undoubtedly increased the expectation of life throughout the world, there is little evidence that it has increased the maximum human lifespan.

The wish to live a long lifespan has also been accompanied by the belief that there may be ways and means of achieving this end. One was the search for the fountain of youth, another was the alchemists fruitless attempts to find the philosopher's stone, or the elixir of life. In more recent times, some have considered the properties of the intestinal flora of carp, which were reported, incorrectly, to live several hundred years. This was the theme exploited by Aldous Huxley in his novel *After Many a Summer.* A more serious claim was made by a famous scientist Elie Metchnikoff, that one of the major causes of ageing was the production of toxins by bacteria in our intestines. He also proposed that yoghurt containing billions of live lactobacilli could neutralise these toxins and extend the lifespan. In support of his view, he cited the long lifespan of inhabitants of the Caucasus, whose diet consistently included yoghurt. Metchnikoff's views lead to an explosion in the sale and consumption of yoghurt in Western countries.

These early claims have been followed by many others in recent times. Anna Aslan, a Romanian physician, believed that procaine, incorporated into a product known as Gerovital, rejuvenated elderly people. In this case her claims were backed up by her treatment of elderly down-and-outs in Bucharest. Once provided with a clean, hygienic environment, and well fed, it was not surprising that they showed rather clear signs of rejuvenation! Aslan's claims were never substantiated, but there has been no shortage of other comparable ones, each with many adherents. One treatment popular in Germany and Switzerland is "cell therapy". This takes several forms, but the basic claim is that extracts of foetal cells of sheep or other animals, when injected into people, has significant rejuvenating effects. A course of treatment is very expensive, and having paid this sum, clients are liable to justify it by saying it was successful. More important, the clinics which provide this treatment have never supplied, from their extensive records, any data showing that their clients had an increased lifespan, in comparison to the population at large. One can be sure that if lifespan extension actually occurred, the facts would be included in the brochures which extoll the beneficial effects of the treatment. There is, of course, absolutely no scientific basis for the claims made.

Indeed, the treatments are potentially dangerous if unknown viruses are present in the cell-free extracts.

Many books have been published which claim to have discovered the means to significantly increase longevity. A typical title might be: *How to live to 100 years*, and there are many others with similar titles. These books appear because publishers know that they will sell very well. Needless to say, the authors of these books do not, on average, live any longer than members of the general population. If any single author did live to a very advanced age, they would probably achieve huge publicity and sales. Some authors have become very prosperous by making huge claims of life extension in their books. One such person is Dr Deepak Chopra in the USA who wrote a best selling book *Ageless Body, Timeless Mind: the Quantum Alternative to Growing Old*. His promise is that illness and even the ageing process can be banished by the power of the mind. He also explains that his views are all firmly grounded in quantum theory. He also combined his writings on well-being and longlife with spiritualism, producing a certain recipe for success in a gullible market. Many years before that George Bernard Shaw wrote a play *Back to Methuselah*, with a very long preface. This contains much biological nonsense, and assertions that one can "will oneself" to postpone death for a very long time. Since he lived 94 years, he did quite well.

Another fruitless quest for immortality is the freezing of dead bodies in liquid nitrogen. There are companies in the USA who claim that if a customer'd body if preserved in this way, the ways and means of bringing the body back to life will be found in the future. Naturally, the freezing and storage of corpses is expensive, so the gullible customers when alive pay out a considerable amount of money, and also have to agree to leave an agreed sum in their will. The result is a considerable profit for the company.

The gullible public, especially in the USA, spends enormous amounts of money on a variety of "life extension" products, sold mainly in healthfood shops or by mail order and the net. As a result of clever marketing, sales are burgeoning. In no case is there any data documenting beneficial effects, but nevertheless the manufacturers and suppliers manage to pursuade the public that such benefits are to be expected. The huge industry that has been built up is supported by innumerable articles in magazines on rejuvenation, regeneration and life extension. These are often backed up statements such as: "Scientists' expect that in the future it will be possible to live to X number of years," where X is usually considerably greater that the maximum recorded lifespan of 122 years.

This brings me to the new movement known as "anti-ageing medicine," and its offshoot that champions the slogan "scientifically engineered negligible senescence" (SENS), the brainchild of Aubrey de Grey in Cambridge, UK. One of the the leading propagandists for anti-ageing medicine, Dr Ronald M. Klatz, wrote in the forward to the book *Advances in Anti-Ageing Medicine*: "Within the next 50 years or so, assuming an individual can avoid becoming the victim of major trauma or homicide, It is entirely possible that he or she will be able to live virtually for ever." Aubrey de Grey has said: "I think the first person to live to 1000 may today be 60 years old."

How is it possible to make these claims? The first requirement is to ignore the huge literature on ageing research. This includes the four books I cited in the Preface, and the fact that scientists who are familiar with the field have discovered the biological reasons for the existence of human ageing. The second is to ignore the enormous amount of information that has been obtained by the study of human age-associated disease. In other words to ignore the many well-documented textbooks on human pathology. The third is to propose that in the future stem cell technology, and other technologies, will allow vulnerable parts of the body to be replaced and/or repaired. The new "bionic" man will therefore escape from ageing.

It took millions of years for humans to evolve to their present body design, which was outlined in Chapter 2. The protagonists of anti-ageing medicine are saying that in a few years that body design can be altered by technologies of "scientific engineering" to a design that no longer ages. So the millions of years of evolution becomes a decade or two. Nothing realistic is said about the way the procedures would be applied to present day people, let alone about the expense that might be involved. The whole movement not only becomes science fiction; it is also breathtakingly arrogant. There are thousands and thousands of reseach scientists carrying out research on age-associated diseases. They are experts and specialists in their own scientific disciplines, with their own journals documenting the latest results, and their own conferences where these results are presented and discussed. The aim of all this biomedical research is to discover the origins or causes of age-associated disease, devise better treatments, and hopefully put in place preventative measures. The anti-ageing fraternity are claiming that they can do better than all these biomedical scientists. They are saying that in a *few years* it will be possible to by-pass all these diseases, and "cure" ageing, as if it is a single disease.

The real reason for the existence of the anti-ageing movement, is the prior existence of the media and a gullible public, that are receptive to the propaganda that is put about by pseudo-scientists. When it comes to life extension, people very much like to be told that they can expect longer lives, especially in the U.S.A. Also, some biogerontologists, who should know better, often justify their work (which may be, for example, on yeast, nematode worms or fruit flies), by saying that it might lead in the future to longer human lifespans. In many cases this also makes it possible for them to attract generous funding for their laboratories. These research scientists should be emphasising the real reasons for studying ageing, which is to better understand the origins and development of many age-associated diseases, so that prevention or cure becomes more successful. The point was well expressed by Alex Comfort, who in his time was a leader in the field of biogerontology:

**The medical importance of work on the nature of ageing lies less in the immediate prospect of spectacular interference with the process of senescence than in the fact that unless we understand old age we cannot treat its diseases or palliate its unpleasantness.**

It is well documented that, unlike many in the USA, people in Europe are more concerned about their health and quality of life in old age, rather than life extension. They are particularly concerned that they may become an increasing burden on their families, and increasingly dependent on geriatric services.

From a scientific point of view it is a worthwhile excercise to consider what would be necessary to produce a complex organism that was potentially immortal. First, the animal would have to be in a steady state, that is, the adult would not be young or old, but remain anatomically and physiologically unchanged indefinitely, in the absence of some lethal accident. Second, it would have the ability to regenerate all components of the body that began to show any signs of ageing or decay. Third, to do this it would have to invest a large proportion of its resources in maintenance, regeneration and renewal. Figure 7 illustrates the principle; it assumes that normal functions are constant, but the remaining resources are divided between reproductive functions and maintenance functions. The immortal animal would have fewer resources for reproduction, fewer offspring, and much reduced Darwinian fitness.

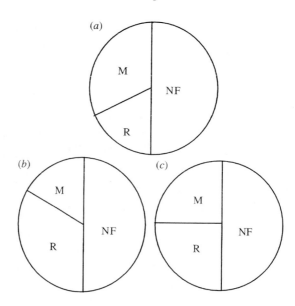

*Figure 7.* Strategies for survival and the allocation of resources. NF are normal functions, which are assumed to be a constant fraction of all available resources. R, all reproductive functions, and M, all maintenance functions. (a) an animal which is potentially immortal; (b) an animal with a short life span, and (c) an animal with a long lifespan

  Maintenance and renewal are expensive, and this is well illustrated by the analogy of a motor car. Everyone knows that most cars have limited lifespan: defects and mechanical problems accumulate until it is no longer economic to repair them. However, a few cars are carefully preserved. These vintage cars are proudly maintained by their owners, but at a cost. Spare parts which are no longer available have to be manufactured at great expense. The cars survive, and may survive indefinitely, but the cumulative and ongoing cost and labour is enormous.

# Chapter 10. Doctors' Dilemma

The continual progress in the quality of health care in developed countries during this century has resulted in an ever-increasing expectation of life. However, the success of medical research, which includes successful diagnosis and treatment, has been achieved at considerable cost. This cost is largely due to the steady rise in age-associated disease in ageing populations. It has been said that in the U.S.A. the costs of medical treatment in the last month or two of life, very often significantly exceed that spent on the whole of the previous life of each patient. In developed countries as a whole, the cost of medical care for individuals over 65 is five or six times higher than for younger individuals, and about 50% of individuals over 65 have some physical disability. WHO statistics show that the costs of health care as a percentage of gross national product in these countries doubled between 1960 and 1986. The cost in real terms will double again early in the 21st century. Much of this expense will be due to the increasing costs of health care from the aged. In addition to medical expenses, general care of the aged consumes the activity of an army of nurses or other health care workers, as well that of many younger relatives.

All this makes depressing reading, and the list of age-associated diseases makes the overall picture even more gloomy. One of the commonest is cardiovascular disease, commonly known just as "heart disease". This is in fact not always a disease, but simply a result of ageing. The heart is a highly efficient pump, but it has limited capacity for repair, and with time it loses efficiency. This may be due to the calcification or other defects in the valves, a loss of muscular activity, or, most commonly, a failure of coronary arteries to provide blood and oxygen. Coronary arteries commonly become blocked through the formation of atherosclerotic plaques. Arteries lost elasticity through the age-related cross-linking of proteins such as collagen and elastin, and this results in a rise in blood pressure (hypertension) and increased risk of stroke from cerebral haemorrhage.

There is strong interdependence between the vascular and renal system. The kidneys receive 25% of cardiac output and about 1700 litres of blood are processed each day. The kidney is sensitive to hypertension, and has itself an active role in the control of blood pressure. Although the kidney is far less sensitive than the brain or heart to tissue damage, its components do lose efficiency with time.

Loss of kidney function is common in old age, and is not uncommon in younger individuals. Also, the regulation of the flow of urine is quite frequently lost in old people, which leads to the problems of incontinence.

Diabetes affects carbohydrate, fat and protein metabolism, primarily as a result of abnormally high glucose in blood and tissues (hyperglycaemia). There are two types of diabetes. Type 1 is early onset and is due to the failure of the pancreas to produce a sufficient quantity of insulin. It is treated by injections of insulin. Type II is less well defined and it is, characteristically, an age-related disease. It is due to abnormalities in insulin metabolism which are not fully understood. Its incidence is greatly increased by obesity. Although it can often be treated by a controlled diet, it is nevertheless a common debilitating disease amongst old people.

Osteoporosis is due to a loss of bone mass, and this commonly causes hip or other fractures in old people. It becomes particularly common in post-menopausal women, but also occurs in elderly men. The addition of calcium to the diet and greater exercise delay the symptoms, but it is nevertheless one of the commonest age-associated diseases. Osteoarthritis is caused by deterioration of the normal structure and function of joints, is frequently associated with painful inflammation. Part of the problem is simply due to the wear and tear of mechanical structures which cannot repair themselves, but autoimmune responses can also be important. This is a general problem in old age as the immune system loses efficiency and the ability to distinguish self from non-self proteins, components of proteins, or other molecules. The decline in the capacity of the immune system to combat invading pathogens can also have severe consequences for old people. It is well known that they become more susceptible to respiratory diseases such as influenza, bronchitis and phneumonia.

One of the most obvious age-related changes is the appearance of skin, particularly on the face and hands, or other exposed parts of the body. One of the major changes is due to the cross-linking and loss of organisation of collagen, which is a major structural component. Wound healing becomes slower during ageing, and in some cases the failure of skin ulcers to heal becomes an important problem. The effects of the ultraviolet components of sunlight are well known. Not only is wrinkling and the formation of "age spots" accelerated, but there is an increasing risk of skin cancer.

The majority of body cancers are known as carcinomas, and it is very well established that the incidence of these is age-related. There is good evidence that the progression of these tumours is a multistep process, so clearly the probability of the emergence of a malignant tumour is time-dependent, and therefore age-associated. The risk of malignancy begins to rise at the age of 50 or so, and thereafter increases steeply to an incidence of 30% or so at the age 80–85. After cardiovascular disease, cancer is the commonest cause of death.

Finally, there is deterioration of brain function with age. Strokes are due to a failure of blood supply to important parts of the brain, which soon suffers damage in the absence of oxygen. This can be due to haemorrhage, often caused by high blood pressure, or it can be due to blockage of an artery by a clot, or a detached atherosclerotic plaque. Neurons are steadily lost as we got older, and this affects memory and other brain functions. Accelerated damage or cell loss gives rise to Alzheimer's disease, which is premature senility. Obviously, the younger an individual is, the greater is the distress caused by Alzheimer's disease, but is the population as a whole, the incidence rises continuously with age, and few very old people are completely free of the symptoms. It is unfortunate that the brain is particularly sensitive to cell and tissue damage, and neurons can never be replaced. The retina is an extension of the brain, and it too suffers damage with ageing. This often leads to poor vision or blindness. Also, the long-lived proteins in the eye lens, become cross-linked, oxidised or have other changes with increasing age, so cataracts become increasingly common. The delicate mechanisms that are essential for hearing lose their efficiency with age, so poor-hearing or deafness becomes one of the commonest problems amongst the elderly.

This long list of age-associated diseases is indeed depressing, but as any physician will tell you, there are many other less well known age-associated diseases. It is not surprising that the overall increase in lifespan results in escalating costs of health care. One reason these health care costs rise is due to the continual development of new sophisticated equipment for diagnosis, and also for treatment. These tools are usually very expensive, but once their worth is established, every hospital has to have one.

In general, physicians are expected to treat each disease as it arises, irrespective of the age of the patient. Unfortunately, in the case of elderly patients, the successful treatment of one disease is all too often followed by the appearance of another. This is a major reason for the ever increasing expense of treating these patients. Whether we like it or not, the law of diminishing returns applies, and this raises serious ethical issues. In practice, a compromise has to be reached so that the younger a patient is, the more appropriate it is to use an expensive medical treatment.

One of the success stories in medical treatment, is the increasing use of surgical repair and replacement of parts. We have long been familiar with dental care, in which cavities in teeth are filled, abscesses can be treated by antibiotics, with or without tooth extraction, and false teeth can be provided as required. Comparable intervention in other parts of the body has been developed much more slowly. However, partially blocked coronary arteries are now commonly treated by by-pass surgery, hip-joints are often replaced with mechanical substitutes and cataracts are replaced with implants. In addition, organ trans-plantation has become common, particularly kidney transplants, but also those of heart or liver. These treatments are expensive, and with regard to organ transplantation, there is an ongoing shortage of donors. Another problem is the need to prevent rejection of foreign tissue, so patients have to take drugs to suppress the immune system for the rest of their lives. This can, of course, have unwanted consequences or side effects.

The general medical strategy to replace more and more parts of the body is providing the maintenance that the body itself is unable to achieve. So in this sense, artificial maintenance is increasing lifespan. However, as a general strategy, it is running into formidable diffi-culties. One problem is the justification of the expense. At what age is it appropriate to adopt surgical intervention? This may depend on the financial resources of the health system itself, or the bank balance of the prospective patient. Obviously, fewer expensive operations are carried out in third world countries than in developed ones. This problem will not go away. Some surgeons believe that the problem of insufficient organ donors will be solved by xenotransplants, that is, the use of organs from animals. Pigs are commonly said to be the most suitable donor animal, because their size and physiology broadly matches that of humans. However, the rejection problem is much more severe with tissue from another species, so at present much research is in progress to circumvent in one way or another, the rejection of pig tissue. Another problem, rarely discussed, is that pigs have a relatively short lifepsan compared to other herbivores of comparable size. One could not expect a pig kidney or pig heart to last much more than a decade, so some surgeons talk about repeated or successive transplants. It is also a fact that transplant surgeons are very well paid for their experience and skills. This financial reward becomes a driving force in itself, particularly in the U.S.A., where many elderly patients are all too ready to pay large medical fees.

Other intervention procedures are often discussed, fuelled by the general hope that as research progresses, more and more difficult operations will be carried out. It is probably time to take a hard look at some of the more fanciful operations which have been aired. One set of future procedures is concerned with the central nervous system. At present severe damage to the spinal cord often leads to loss of limb function. Although, minor nerve can regenerate to some extent, major nerves do not. This has lead to much discussion of the possibility of nerve regeneration in the future. However, it is not only a question of stimulating the growth of new nervous tissue, there is also the even more important problem of the specificity of nerve connections. Successful repair of a spinal cord involves perfect matching and rejoining of the many thousands of severed nerve endings. The enormity of the problem can be best illustrated by a consideration of the likelihood of eye transplantation in the future. Every sensitive cell in the retina is connected to a corresponding cell or cells in a specific part of the brain, known as the visual cortex. All these nerve connections, or axons, pass along the optic nerve, and there are about a million of them. After cutting the optic nerve in an eye transplant operation, all these specific axon endings must be rejoined. It is not just a question of simple regeneration or repair, but of one where complete specificity between retina and brain is retained. At present, it is not even understood how such specificity is laid down in the early development of the eye. (Although some preliminary understanding has come from the study of experimental animals, such as the toad Xenopus, which has much greater powers of nerve regeneration than mammals have). This is an area where science fiction enters the province of real medical or biological science.

The problems of medical treatment are not just associated with advances in surgery, or other examples of high-technology medicine. Let us consider again the list of age-associated diseases. At present, considerable resources are being devoted to research on everyone of these in laboratories all round the world. There are specialised clinical and non-clinical research scientists studying heart disease, diabetes, osteoporosis, Alzheimer's disease, cancer, and so on and so on. There are specialised journals which publish the results obtained. There are national and international conferences, where experts in each disease exchange the latest information. There are huge pharmaceutical companies competing with each other, often carrying out their own classified research, with the aim of finding new drugs for better treatments. The aims of all this research are well known, first, to find better treatments of each disease; second, to prevent or delay the onset of the disease, and third, to understand why the disease arises in the first place. These aims are completely laudable, and everyone would wish success for the research teams involved. What is unfortunate is that those studying one disease, for example, late-onset diabetes, do not communicate with those studying another age-related disease, such as cancer. Those working on cardiovascular disease do not talk to those studying osteoporosis, and so on. The study of each disease is at present regarded as a complete discipline in its own right.

The truth is that the second and third aims of the research, that is, the need to understand the origin of each disease, and the need to devise means of prevention or delay, should be within the province of ageing research itself. We saw in earlier chapters that the physiology and anatomy of the body is designed for finite, not infinite, survival. We also saw that maintenance of body functions comprises a set of mechanisms, which taken together use a considerable proportion of body resources. Maintenance is never perfect, so defects arise in individual components of cells, in DNA, in proteins, in membranes, and so on. These defects in cells also affect tissue and organ function in all parts of the body. Although at the tissue and organ level ageing may have multiple causes, at the cellular level there are common features to do with the types of damage or imperfection that can arise in the molecular components which are common to all cells. Thus, there are changes in proteins, such as cross-linking, which may have important deleterious effects in arteries, in the brain, in the lens, in joints and skin, and so on. Therefore all those scientists studying degenerative age-related diseases should not only have a real interest in those fundamental molecular changes, but they should also share and exchange information about the origin of all the different diseases. In a word, they should be interested in ageing *per se*. This is certainly not the case at the present time. Moreover, the amount of resources devoted to research on ageing, especially in Europe, is quite small. Ageing is all too often side-lined as a pseudo-scientific field, together with the belief that those working in it are only trying to increase the maximum lifespan. The truth is that all those scientists studying age-associated disease are themselves studying ageing, even though they may be unaware of this.

Unfortunately there is little sign that the general message is getting through to the medical community. In 1942 the geratrician Edward Stieglitz wrote an article on "The social urgency of research on ageing". The following passage is particularly apposite:

**The shifting age distribution of the population with its increasing proportion of those in the older age groups has introduced innumerable problems of the most practical and significant character. The situation is without precedent. The millions of elderly are here. There will be more. Many are well and capable of continued productive and creative effort if given opportunity to work within their capacities. Others, and there are many, are prematurely disabled by the insidious chronic and progressive disorders so frequent during the senescent period and become heavy burdens upon the family and society by reason of the long course of their disablement. Gerontology, the science of ageing, is divided into three major categories: (a) the problems of the biology of senescence, (b) the clinical problems of ageing man, and (c) the socio-economic problems of ageing mankind. These three categories are intimately and inseparably related; progress in one field us dependent upon progress in the others and vice versa. Clinical medicine, in its youngest speciality, geriatrics, can do much to solve the most distressing problems in the social field. To name but two potential contributions of vital importance: Personal preventive medicine may greatly reduce the toll of premature disability and bettered diagnostic methods can more clearly define the limitations which go with normal ageing. Yet even more intimate is the dependence of clinical medicine upon advances in the fundamental sciences in elucidating just what ageing is, what it does, what retards or accelerates it and why. *It cannot be over-emphasized that the more we know about the biologic mechanisms of the ageing processes, the more effectively can clinical medicine treat the ageing and the aged* (italics added).**

If this was the case more than 60 years ago, it is even more true and relevant today. This is the doctors' dilemma. They are trained to treat each disease as it arises as best they can. The general physician has a broad knowledge, who refers patients to specialists with detailed knowledge of a particular disease. These consultants read the appropriate literature to keep up with current research in that particular field. The whole medical scene has become compartmentalised, and every group of specialised physicians is promoting its own interests, including the provision of expensive equipment. This is the situation which leads to escalating health costs, especially in developed countries, and to costs that certainly cannot be afforded in developing countries. Instead of developing more and more expensive treatments for age-associated diseases, the emphasis needs to be altered

---

Major age-associated pathologies

---

        Cardiovascular disease
        Cerebrovascular disease (stroke)
        Dementias
        Cancers
        Late onset diabetes
        Kidney failure
        Osteoarthritis
        Osteoporosis
        Cataracts and retinopathy
        Loss of hearing

---

Some non-pathological changes during ageing

---

        Wrinkling and sagging of skin
        Whitening of hair
        Loss of muscle strength
        Development of pigmented age-spots
        Loss of lens accomodation
        Tooth decay
        Incontinence

---

so that more consideration is given to prevention of disease. And to be successful in prevention, one needs to know in detail the reasons for the origin of each disease. Intervention at this stage not only saves money, but also improves the quality of life of old people, a theme which will be explored in the next chapter.

# Chapter 11. The Modulation of Longevity

Chapter 9 discussed the mythical claims, ancient and modern, that humans can have very long lifespans. This one is devoted to the various ways in which ageing and longevity can be modulated, including either an increase or a decrease. We know that there are at least three major components that determine the expectation of life at birth. One is the species itself, because in mammals maximum lifespan varies by at least 30-fold, and if we accept the reports of extreme longevity in some whales, probably 60-fold. The second is the environment, which can obviously limit natural lifespan through predation, starvation and disease. Even in a protected environment of a zoo, the environment, including food, and possibly the opportunities for normal well-being and behaviour, are likely to influence longevity. The third, is an intrinsic variation of the determinants of longevity within the animal itself. We saw in Chapter 1 that identical twins do not have have identical lifespans, and even more telling are the inbred mice which have the same genes and live in the same environment. Their lifespans vary considerably, showing that chance or random events are very important determants. It would therefore be surprising if many ways and means were uncovered that had clear measurable effects on longevity.

One of the best examples is the effect of food deprivation on lifespan. It was discovered many decades ago that a low calorie diet significantly increases the lifespan of mice and rats. When animals receive only about 60% of what they would normally consume, their lifespan is increased by 40–50%. Moreover, the significant pathological changes that are normally seen during senescence are very significantly delayed. Another strong effect is on reproduction, because calorie-restricted rodents lose fertility, and may become sterile. This provides a strong biological clue as to why reduced food increases lifespan. In natural environments, small ground living animals are very likely to encounter a variable supply of food. A food glut allows continual reproduction, and populations may rapidly increase. In contrast, a shortage of food makes reproduction very difficult, and young animals may die from malnutrician. It is now widely agreed that the increased longevity of calorie-restricted animals is an evolutionary adaptation. The adaptation is cessation of breeding, and instead that resources available are chanelled into maintaining the adult in a healthy state. When the food supply increases, reproduction begins again. It has been established experimentally that calore-restricted females, when provided with a normal diet, can breed at later ages than animals kept throughout on a full diet.

An important question is whether calorie restriction increases lifespan in other mammalian species, and particularly larger species with much longer longevities than rodents, such as primates and humans. Experiment with monkeys are underway, but they take a very long time to complete, and the results so far cannot be regarded as conclusive. There are those who believe that large slow breeding mammals that have a fairly constant diet, such as tropical monkeys, will not respond to a low calorie one in the same way as rodents. However, we know that human females who are very physically active, such as ballet dancers and some athletes, may cease to menstruate. So in these cases the diversion of much available energy to physical activity, reduces fertility. It might be possible to obtain epidemio-logical data about human diets and longevities, but so far no definitive information has been obtained. There is a large literature that discusses the possibility of substantial increases in human lifespan by reducing calorie input. In the whole of this literature is is rarely mentioned that calorie-restriction has strong effects on fertility and presumably libido.

Another well documented example of increased lifespan was obtained by the Australian gerontologist Arthur Everitt. He removed the anteria pituitary of young rats (hypophysectomy) and discovered that this had an effect on lifespan comparable to calorie-resticted animals. There have been many studies of hormones on ageing, and claims that hormone treatment increase longevity. One is human growth hormone, because elderly males report increased muscular strength or other rejuvenating effects after a course of treatment. Another is melatonin, produced by the pineal gland in the brain, which is important for the control of sleep rythms. A number of extravagent claims have been made about about its beneficial effects in preventing age-related disease, or increasing the longevity of mice. The hormone DHEA (dihydroepiandrosterine) has also been credited with strong anti-ageing effects. There is no doubt that experiments with hormones will continue for a long time to come, and it is likely that modulating effects on longevity will be uncovered. However, the extreme effects discussed in Chapter 9 are very unlikely to be seen.

Antioxidants have become particularly popular, especially in the U.S.A., and these include Vitamin E and Vitamin C, and also carotene. Their efficacy is related to the strong view, which in some quarters is a belief, that the major cause of ageing is the damage produced by oxygen free radicals. If this is so, then antioxidants should be effective in reducing this damage, a conclusion that has been widely accepted by the public at large. The facts are not always easy to obtain. Lifespan experiments have been done with experimental animals provided with a diet containing one or more antioxidants in comparison to untreated control animals. Early reports suggested that life-extension was achieved, but those have not been repeated. A recent experiment carried our in Alec Morley's laboratory in Adelaide, demonstrated absolutely no effect of vitamin E on the longevity of mice. It is a reasonable presumption that other experiments that gave negative results have never been published, and remain unknown. One can, in fact, be fairly sure that any experiment that gave a positive result would not only be published, but would also be given extensive coverage in magazines and newspaper articles.

A detailed long-term study was carried out in Finland with physicians who supplemented their diet with Vitamin E, or Vitamin C, or both, or neither. Their health was carefully monitored over a number of years, and the incidence of age-associated maladies or disease was properly monitored. There was no evidence that either anti-oxidant had any rejuvenating effects, nor significantly altered the normal incidence of age-related pathologies. (There was one possible exception among the twenty or more studied, but note that routine statistical tests of significance give, by chance, a significant result once in 20 tests). Not surprisingly, the result of this important study has not received much publicity.

An interesting suggestion has been made by the gerontologist Roy Walford. In cold-blooded animals it is well known that lifespan is related to body temperature, that is, the lower the temperature the longer the lifespan. Walford pointed out that this might apply to warm blooded animals as well. If body temperature was reduced to 34° C or 35° C, lifespan might be very significantly longer. Although there are means to change body temperature within certain limits, obviously the proposal cannot be tested directly. However, an indirect test exists. Human cells in culture have a well defined lifespan, so in my laboratory years ago we decided to look at the effect of incubation temperature. It was found that cells grown at 34° C had exactly the same lifespan as those grown at the normal temperature of 37° C. On the other hand cells at 40° C had a sharply reduced lifespan. This experimental system has been exploited in other ways to try to uncover the causes of cellular senescence. Many years ago it was reported that the addition of Vitamin E to the culture medium doubled the lifespan of these cells. Unfortunately this claim had to be retracted, because the result could never be repeated.

Another more recent study in my laboratory was carried out with a small naturally occurring peptide known as carnosine. This is present at high concentration in tissues such as brain and muscle. Carnosine consists of an unusual amino acid not found in proteins called (Beta alanine, linked to one of the normal amino acids, histidine). It was found that human cells grown in the presence of a physiological concentration of carnosine had a significantly increased lifespan. Moreover, the cells retained a juvenile appearance throughout growth, whereas normal cells develop a number of obvious abnormalities when they become senescent.

Senescent cells switched to a medium containng carnosine were rejuvenated. Such rejuvenated cells when deprived of carnosine quickly become senescent. The cells used are called fibroblasts, which secrete collagen and are an important component of skin tissue. Carnosine has been marketed under the trade name (Beta-alistine), for the prevention of skin ageing. It is possible that the peptide has an important role in cell maintenance, and it is highly significant that the amount in human muscle tissue is about twenty times higher than that in mouse tissue. Another treatment has also been discovered which has the effect of preventing the usual features of senescence in long-term cultures of human cells. This is a well known plant hormone kinetin, and it is effective at very low concentrations. There is no information about the effects of carnosine or kinetin on the longevities of experimental animals.

If any new treatment or chemical is reported to increase longevity, it receives wide publicity. Even ageing experiments in yeast (where the finite number of new cells budded off a mother cells is counted), are extrapolated to extravagent claims that the same treatmnent would increase human longevity. This was the case for resveritol, which is present in red wine and some other plants or their products. It was reported to increase life spans in several different experimental systems. However, other laboratories have not seen the same effects.

Another approach is known hormesis, which is the response to repeated treatments that induce mild stress. Such stress can be induced by heat shock, pro-oxidants, heavy metals and irradiation, and it is reported that the overall regime of treatments, has measurable beneficial effects in several experimental systems. The interpretation is that cells that are induced to protect themselves from stress (which is a well known phenomenon), also turn on cell maintenance mechanisms that are beneficial. Extrapolating to humans is a difficult question, because repeated mild doses of irradiation, or low levels of heavy metals, would hardly receive medical approval, especially as any beneficial outcomes might take years to be seen.

One of the documented effects of irradiation is in fact a shortening of the lifespan. There were many studies done after the second world war, and the earlier senescence of irradiated animals was very well documented. There is now little discussion of these results, and some claim that the induced premature ageing was different from normal ageing. However, in the most thorough studies, post-mortem examinations showed that there was little, if any, difference in causes of death in the treated and control animals. A major target for irradiation in cells is DNA, and it is likely that DNA damage is responsible for the shortened life span. The important damage may be of a type that evades normal repair, but to this day it has not been identified.

There is now huge interest in life style effects on disease, general well-being, and that part of the whole lifespan that is sometimes called the healthspan. In addition to the expectation of life at birth, there is now a recognised statistic known as the expectation of a healthy life. Obviously it is the aim of gerontologists and geriatricians to bring the expectation of a healthy life as close as possible to the total expectation. With early diagnosis and increasingly successful treatments this aim is to some extent being achieved. Much more needs to be learned, for example, about the long term effects of excercise, or of particular diets on the healthspan. Unfortunately, much is also being learned about the long terms effects of unhealthy diets. Obesity has become something of an epidemic in some developed countries, and particularly in some communities in the USA. This is associated with the appearance of late onset diabetes, which has many deleterious side effects that together bring about a reduction in lifespan. This is a major cause of the shorter expectation of life of aboriginal people in Australia, nearly 20 years in comparison to the rest of the population. The primary cause is probability the availability of cheap carbohydrate foods, coupled with a scarcity of money. One would have thought that this problem would be solvable.

There is a great deal to be learned about lifesyle and longevity. One way to obtain information over a long time scale is known as a longitudinal study. The most important of these was started nearly fifty years on the initiative of Nathan Shock, and it is known the Baltimore Longitudinal Study of Ageing. Healthy volunteers from all age groups are enrolled, and then examined and tested every two years, using over 100 procedures. The study documents life-style parameters, "biomarkers" of ageing, health and illness for a substantial number of individuals, from the time they volunteer to the time of their death. Not as much information has come out of the study as one might have hoped, but presumably there will be definitive reports and publications.

Western developed coutries have grown used to a gradual increase in the expectation of life at birth, and with the increasing success in the treatment of age-associated disease, this trend will continue. Yet the total elimination of carcinomas as a cause of death, the expectation of life would not be greatly increased. The main reason for this is that there are multiple causes of ageing and the elimination of one does not affect the other. So if you do not die from cancer, you may die from a heart attack, from a stroke, from kidney failure, and so on. It has been calculated that if all major diseases were eliminated the expectation of life would increase by about 15 years. There are still remain less important age-associated diseases, and a multiplicity of cellular and molecular defects accumulating all the time. Apart from everything else, the treatment of every new pathology as it arises becomes prohibitively expensive. It is yet another example of the law of diminishing returns – the more expensive the treatment becomes the smaller the gain in terms of healthy life is gained.

# Chapter 12. Ageing and the Angels

Amongst all animal species, human beings are unique in knowing that they will eventually die. Even young children are aware that elderly relatives do not have long to live, and that when they die they will not see them again. This realization has had enormous cultural and social consequences.

Nowadays, individuals are expected to reach an advanced age, but this was not the case during the early evolution of man. Annual mortality was such that few people reached natural old age. We saw in Chapter 8 that in early human populations the expectation of life at birth is less than 20 years, and about 3% of individuals would be expected to reach the age of 45 years. In a society such as this, death would be a common phenomenon, but these deaths would be predominantly due to starvation, disease or predators. Death from "natural old age" would be a rare event.

It is probable that these hunter-gatherer societies consisted of an extended family, or a few families living together. The community was very important in food gathering, hunting and also in protecting itself against danger. As communication skills developed, children would be taught the importance of altruistic behaviour. Such behaviour increased the changes of survival and reduced the likelihood of death. In contrast, selfish or acquisitive behaviours would, at best, only have a short term advantage in increasing survival. Under these circumstances, it is fairly easy to see that a basic moral code would develop. To increase the chances of one's own and one's relative's survival, it became important to contribute to beneficial group activities, such as food gathering, hunting and defence. Since the groups contained many genetically related individuals, kin survival was an important component of group behaviour, and this of course resulted in kin-selection.

It is fairly easy to envisage the form this moral teaching might take. It could be one of the roles of the more experienced members of the community, as well as being undertaken by the parents of children. It would be evident to all that the environment was overtly, or potentially, hostile and stressful. The benefits of successful hunting and the hard work of food gathering would be explained, as would the hazards of predators. Also, the danger of being alone would be stressed, since the safety and survival of community depended in very large part on co-operative activities. The danger of the unknown in a hostile environment would be very important. This is in part built into our sensory system, since a substantial part of the retina is particularly sensitive to movement in our peripheral vision, which is just what is needed to detect danger approaching. It would not be surprising if the teachers in the society warned of unknown enemies, evil spirits and so on, which everyone must be made aware of. Such dangerous imaginary beings may well have been the first non-material creations in those early human societies.

The moral code was important for survival, because those who followed the teachings would be much more likely to survive than those that did not. In those early societies, the major reward would be a relatively long life, with the opportunity to raise a family oneself. Old age and natural death would still be an uncommon, possibly a very uncommon, event. Indeed, it is possible that members of such communities were taught that they might survive indefinitely, if they were skilled and also lucky enough to avoid the many hazards inherent in their environment and life-style.

This period of human pre-history lasted a long time, with populations remaining fairly small. As time went on, human skills gradually improved, particularly in communication and tool-making. This lead to a somewhat lower annual mortality, a reduction in risk level, and as a consequence a gradual increase in population size. This provided the driving force for migration from Africa to Europe and Asia, and many believe there were successive waves of such migration, involving various hominid sub-species.

The success of human adaptation to the environment increased inexorably, and eventually lead to the means to control the environment itself. This was seen particularly in the development of agriculture. Instead of a nomadic existence, or one based on habitation in caves, humans began to plant, tend and harvest crops, at least for one part of the year. The advantage, most obviously, was a more reliable food supply. Associated changes would be the building of semi-permanent or permanent shelters, and an increased size of each community. In addition, there would be division of labour within each community, for example, tool-makers, farmers and hunters, albeit no doubt with much overlap between them.

These trends would have had very profound effects on the existing moral codes of behaviour. For the first time, the reduced mortality would result in the survival of some individuals to old age. The reward for hard-work and altruistic behaviour would not be indefinite survival, as the community elders had previously taught, but senility and decrepitude. Individuals who survived into old age would feel cheated, because it became all too clear that adult life could not be prolonged by altruism and community spirit. This, I believe, provides the key to the origin of one of the commonest features of human religion. The social solution to the problem of senescence, old age and death, is simply to invoke an afterlife. The existing moral codes would remain in place, but the eventual reward for virtue would be changed. Instead of the benefits of an increased likelihood of survival in a normal human community, the emphasis would be shifted to the benefits of survival in a non-material afterworld. In this context, immortality, paradise, reunion with long deceased relatives or friends become added incentives. Once the concept of an afterlife came into being, it would also be possible to relate the quality of that afterlife, to behaviour in the real world. Thus, individuals who followed the accepted moral codes would be rewarded with a favourable afterlife, whereas those who did not have the appropriate community spirit, and instead adapted selfish behaviour, would be faced with the prospect of a hellish afterlife.

This does not mean that one or more individuals consciously devised a social solution to the "problem" of old age. It would have been gradually realised that the strength of the moral teaching was being undermined by the dire consequences of growing old. It would have been simple wish-fulfillment to invoke other worlds inhabited by the souls or spirits of mortals. It is obvious from the contemporary world that people continually draw comfort from their belief in an afterlife. It would have been so many thousands of years ago as well, where humans had already evolved most of their intellectual and emotional capacities. Those who first suggested the possibility of an afterworld wanted to believe in it themselves. Religious faith was born. Individuals who promulgated this new faith also believed they had God-given powers, and their success in persuading others that these powers were real could over-ride normal experience and expectations. Most religious prophets back up their teaching and authority by the claim that they are in direct contact with an all-powerful deity.

In this new context, the child became much more aware of death following old age. The answer to the question "why do we die?" can be dressed up within the context of any of several supernatural God-given or God-driven worlds. Previously, children were told that relatives died because they got ill, were killed by predators, or starved. Indeed, if they were spared all these causes of death, and others, they might well live on this earth forever. This would have been a great comfort to them, but one, unfortunately, that became demonstrably false. It was necessary for parents and elders to provide new answers to the child's question.

Speculation about the exact form of any particular form of early religion is not very helpful. Social anthropologists have documented the diverse beliefs of many different cultures. From a sociobiological standpoint, it is sufficient to define the basic characteristics of a religion which had a clear social function. First, the religion would be based on a set of beliefs, formulated as dogmas that were not to be questioned. These would be a uniform set of dogmas with any one community and it would also be transmitted from generation to generation. Second, there would be a specific function for priests or priest-like members of society. They would be responsible for teaching the beliefs to the rest of the community, and they would be endowed with authority by being in contact with whatever Gods were invoked. Third, the existence of omnipotent deities provides a framework which answers the questions any conscious being might ask. Who created the world or universe? Who created man? Why do men and women grow old and die? In all cases the answer is the deity, who is responsible for creation and defining the course of life and death. Although a belief in human "free-will" (for example, the ability to choose between good and evil) may be a component of religion, a strong belief is an ordained predetermined fate is also very common. Fourth, the religion would be closely related to the moral values that had probably predated the religion itself. Thus, worthy altruistic behaviours would be rewarded, and acquisitive, selfish or antisocial behaviour would be punished. This punishment might occur within the society itself, or it might become associated with the threat of a hellish afterlife, or re-incarnation to a lesser species. The invoking of evil spirits, unknown dangers and so on, might well be related to this aspect of religion, in order to deter antisocial behaviour. Fifth, the whole religious edifice would be built around a series of myths, about the origin of gods and humans, as well as visual symbolism. Instantly recognisable images of gods, or other mythical figures, would be essential components in gaining social acceptance, as would ceremonies and ceremonial events with individuals in elaborate dress. Many of these activities became intimately associated with art and music, and particularly, later on, with architecture.

These five features of early religious faith are not at all isolated one from the other. There would be much overlap and interaction between them. The links between mortal beings and immortal gods would be variable between different religions. Mortals can become god-like, and gods can enter normal communities. We know the potential power of one man, the prophet, is enormous. Yet this occurrence is likely to be rare, and for every successful prophet who founded a whole religion, there are probably thousands of cult figures who would themselves like to found their own particular religion by impressing their own beliefs on others. The complexity of modern contemporary society does not appear to have reduced the frequency of cults. Successful religion must have permanence in a particular community, or set of communities. Richard Dawkins in his book *The Selfish Gene*, used the word "meme" to describe a belief, concept or idea (among other things), which is transmitted from generation to generation, and he cited Judaistic memes as examples of some of the longest surviving ones. However, the time scale of memes is minute compared to the time scale of genes.

In this Chapter and Chapter 8, I have attempted to explain both the evolution of human longevity, and the social consequences of the increased awareness of natural death from old age. Darwinism successfully challenged the authority of contemporary religions. Although sociobiologists such as Edward Wilson have provided speculations about the evolution of human religions, I believe that one can go further and provide a much more rational explanation of origins of religion.

Religion has been the dominant force over most of human social evolution, but only at the end of the twentieth century does it become possible to understand why this was the case. We no longer need mythical explanations for the inevitability of old age and death, because we now have rational scientific ones. Of course, only a minute fraction of humanity at present has this knowledge and awareness, and most will continue with beliefs little altered from the past. Those that reject these ancient views have accepted the scientific facts, namely, that human individuals on this planet have finite survival time and that there is no non-material afterlife. This realistic view of ourselves and the world we live in is not just a rejection of faiths, but encompasses the very strong belief that all human problems must be solved by human beings themselves. Human beings alone have knowledge and reason, and must act accordingly in the environment they find themselves, to face up to and solve their own problems. A belief in Gods, angels or other mythical beings, as well as a belief in fate, will simply make these problems harder to solve in the end.

# Chapter 13. Longevity, Population Pressure and Warfare

The emergence of man from his pre-hominid ancestors lead to the formation of hunter-gatherer communities. The scarcity of food always limited population growth and also population density. For example, it is estimated that the indigenous population of the huge land mass of Australia when the the colonisers arrived was probably no more than half a million. It can be calculated that the rate of population growth from a founder population about 50,000 years earlier was less than 0.3 per cent per year. This must also mean that the expectation of life at birth was significantly less than 20 years. If it was any higher than this a much greater rate of population increase would have occurred.

If there is a defining moment in human social evolution, it must have been the development of agriculture, which is thought to have originated about 10,000 years ago. It would not have happened in one step, but in several successive steps. The impact of growing plants for food was enormous. It meant that communities became established in the place they cultivated crops. It meant also that food could be stored and eaten whilst new crops were sown and grown. Wells were dug to obtain water, or communities settled near rivers or lakes. The domestication of animals was equally important, since they provided a source of meat and milk. Probably some time later on, they helped till the land.

The largest effect of agriculture and a steady food supply was on population size and density. This major demographic change is the result of an increase in expectation of life at birth. As soon as this average longevity reaches about 20 years, the population starts to rise. Families would become larger, and many more children would survive to become reproducing adults. If the number of offspring increases by 10%, per generation, then the population rises exponentially, and will double in less than eight generations. More females would reach the end of menaupause, stop reproducing and care for grandchildren or other family members. In these settled communities there was no longer the driving force of migration, which was common during the hunter – gatherer period.

Instead, more huts or houses were built; more land was culti-vated, and more domestic animals were bred. Tiny villages became larger villages, and then towns, so he population density continued to increase. The social structure of communities changed, for example, there was division of labour. Although farming would be the main activity, the making of new tools and the building of houses would demand special skills. There would be a need for pots for cooking, and the weaving of wool or other thread for clothes. As we saw in the previous Chapter, the simple altruistic kin selection of hunter-gatherers would be replaced with a set of moral values that could be applied to whole communities.

Many animals are territorial, and this can lead to skirmishes of various kinds. Human hunter-gatherers would also have territorial claims. particularly to protect known food supplies. However, because population density was alway low, conflicts would necessarily have been on a relatively small scale. Individuals might have sometimes been killed, but communities living at subsistence level could ill afford to lose adult males. Perhaps a more common type of skirmish would arise from raids in which adult or adolescent females were captured or kidnapped and taken to live with the raiding group or tribe. This would have the effect of increasing the number of births, and therefore the survival of that tribe at the expense of a depleted neighbour.

There are two interesting models of early warfare which occurred on a limited scale. In Polynesia, human colonisers had seafaring canoes and agriculture. There were plentiful supplies of fish, and also of agricultural products, including meat from domesticated pigs. The main shortage was simply living space, so that as the population size increased, there was increased stress due to the limitation of arable land, and a general reduction in available food per head. This stress would lead to competition and conflict, and the upshot would be that part of the population would be forced to leave in canoes to find new living space. This could only be done by searching for uninhabited islands. Thus the whole of Polynesia became colonised, apart from the smallest islands.

The second model is seen in the highlands of Papua New Guinea. These tribes also have agriculture, but for thousands of years had only stone and wooden tools and weapons. The amount of land which can be cultivated is very limited, and is also essential for continued survival in very restricted living conditions. In this environment, defence of territory becomes very important, and if some can be gained from a neighbouring tribe so much the better. Tribal warfare became in effect a way of life. Much of this warfare must have been, and probably still is, threat and bluster, rather the real violence which results in much loss of life, although head hunting amongst enemies certainly occurred. We see the emergence of warriors, who spend a lot of time making weapons and shields, and take a pride in their exploits. In this kind of human warfare we find for the first time the concept of a dual morality, namely, the morality within a tribal group which is based on altruism, family values and a shared need for food and shelter. The neighbouring tribe is subjected to a quite different morality, where aggression and the used of weapons, such as spears, arrows, clubs and stone-headed axes, became an essential part of human behaviour. The origin of this continual warlike state cames from the fact that every habitable valley becomes colonised, but living space is very limited. Ultimately, it is derived from human success in mastering poor habitats, increasing survival and longevity, and thereby becoming too overcrowded.

Once human populations start to increase in size, as a result of increased average longevity, the rate of increase will be exponential, as was first clearly pointed out by Thomas Malthus at the end of the eighteenth century. Nevertheless this rate of increase can continue for long periods during human social evolution. Villages become small towns which turn into large towns. Large towns become small city states, or nations, supported by an extensive agricultural base. Technologies were accelerated in many ways, with more imaginative use of wood, weaving and pottery, and there were larger, more elaborate buildings. Later there was the smelting of metals: first copper and bronze, and then iron and steel. City states eventually turned into empires with very large populations. It is no accident that two of the first great empires were on the banks of great rivers: Egypt next to the Nile, and Mesopotamia on the Tigris and Euphrates. There was a plentiful supply of water for the agriculture essential to support very large populations.

These civilisations became very sophisticated, with large palaces. temples and monuments, arts and crafts, and written languages. They also had complex social structures, including kings and governments. It was probably at this time that the human activity of organised warfare became an accepted part of civilisation; weapons and armour were manufactured in great quantity, and horses were trained. The wheel had been invented, so chariots could be used. Above all, there was plenty of manpower, simply because the populations had grown so large. There were many reasons for warfare: acquisition of much needed land; the capture of resources and treasure; the overthrow of an oppposing ruler or king, or a clash between different ideologies or religious beliefs. Armies and battles became larger and larger, as did the number of people killed. Cities and fortresses were plundered or raised to the ground, women were raped, and civilian populations often put to the sword. All this became an acceptable part of human behaviour, namely, the dual morality of peace among law-biding citizens, and violence against foreign populations elsewhere. Not only that, but also the widespread glorification of warfare, with the lauding of victorious military heroes, and the honour and respect of those that fell in battle, apart from, of course, the enemy dead. What began thousands of years BC remains standard practice in the 20th and 21st centuries AD. Throughout this time the weaponry of warfare has become more lethal, and far more expensive. This has been accepted by most of the public at large.

All this became possible because there were so many people that could be involved in warfare. Not only could very large armies be assembled for battle, but the loss of thousands of soldiers and civilians could be readily sustained by populations that had enormous reproductive potential. Huge loss of life in one war could be followed a generation later, as happened in 1914 and 1939. It will rarely be admitted that life is cheap, but the reality is that in human populations over thousands of years, civilisations have accepted death in war as a normal part of the human condition. One of the major paradoxes of human behaviour is that murder in a normal domestic environment is one of the worst crimes, only exceeded in many cases by treason against one's own country.

At the end of the nineteenth century, Herbert Spencer well understood the different moralities that are applied to peaceful populations and to warfare. He wrote:

**Rude tribes. . . . and civilised societies . . . have had continually to carry on external self-defence and internal co-operation – external antagonism and internal friendship. Hence their members have acquired two different sets of sentiments and ideas, adjusted to two two kinds of activity. . . . A life of constant external emnity generates a code in which aggression, conquest and revenge, are inculcated . . . . . . . . . Conversely a life of settled internal amnity generates a code inculcating the virtues conducing to harmonious co-operation.**

These two different sets of sentiments and ideas he called the "code of amity" "and the 'code of emnity."

In none of the many published discussions of the origins and reasons for human warfare has an increase in human life expectancy at birth been mentioned as a major cause. Nevertheless, the adaptation of humans to their environment, their reduced mortality and their social evolution over a few thousand years, significantly inceased average lifespans. The continued demographic trend of increased longevity and population size became unstoppable, and ultimately gave rise to human warfare.

# Chapter 14. Dialogue between Life and Death

At first sight there seems to be little relationship between the origin of life on this earth, and the ageing of its most advanced organisms. Surprisingly, there are connections, which constitute part of the paradox of life. Over an enormous time scale, molecular evolution based on natural selection produced the first living cells, and once these appeared, more and more complex organisms evolved, again over an extremely long period of time. Life had triumphed over the non-living inorganic and organic world. Nevertheless, the most complex living forms, the higher animals, are unable to sustain this life. Over a very short period of time, in comparison to evolutionary time, the complex structures of the body loses their capacity to preserve themselves, and the body then ages and soon dies. The molecules which were previously assembled into beautifully integrated functional structures, revert to disorganized components, which, in a natural environment, simply become nutrients for lower organisms. In this brief Chapter I explore the dialogue between life and death during evolution.

Most scientists agree that prior to the origin of life, a variety of organic molecules would accumulate in aqueous environments. This is the so-called "primeval soup", the constituents of which came from the action of radiation and lightning on simpler molecular components. Many of these reactions have been demonstrated under laboratory conditions. The molecules accumulated over very long periods of time – because they were often chemically stable, and there were no microorganisms to break them down to simpler components. (Nowadays any mixture of organic molecules, such as that the primeval soup, would simply provide food for the many bacteria, or other microorganisms, which can extract energy from these molecules by degrading them).

There is much disagreement about the way small organic components might have assembled into larger molecules, the replication of such molecules, and their eventual organisation into simple single-celled organisms. Nevertheless, there are some basic features of primitive organisms about which there is agreement. The genetic material consisted of nucleic acids, probably RNA in the first place. The first catalytic enzymes also probably consisted of RNA, as some of the crucial features of such catalysis have now been discovered in laboratories. Proteins may well have appeared later, with their structures coded for by RNA. The present day genetic code is the same from bacteria to man, so it is very likely that there was only one origin to the genetic code. Also, almost all the individual components of proteins, the amino acids, can exist in two forms either with a right-handed structure, or a left-handed structure, with identical atomic components. (When amino acids are synthesised chemically, they consist of equal mixtures of the right and left-handed forms). All proteins today contain only left-handed amino acids, which again strongly indicates that there was a single origin of living organisms.

Classical physics tells us that large numbers of molecules will always achieve a steady state of maximum disorder. It is a law of physics, but it is violated by the evolution of living organisms. The availability of energy is an essential ingredient for the formation of ordered structures capable of replication, and this energy could have come from light, from organic molecules in the primeval soup, or in some cases from inorganic chemicals. These primitive organisms, and all their more complex descendents, were based on the properties of rather few types of atoms, namely, carbon, nitrogen, oxygen, hydrogen, phosphorus and sulphur, some common salts and trace amounts of certain metals. The ingredients were much simpler than the wide range of atoms and their compounds in the non-organic world. Nevertheless, the simple basic ingredients were gradually built up into the ever increasing complexity of the organic world.

These early organisms would have contained nucleic acid molecules (the primitive genes) capable of replicating themselves, proteins and probably membranes, which enclosed all the constituents. There would have been many mistakes in the replication of genes and the correct assembly of proteins. This probably meant that reproduction was followed in many cases by death of the organism, because the molecular order which is essential for life could not be sustained. Obviously, in early evolution accuracy would be an advantage, so natural selection gradually increased the precision of gene replication and protein synthesis. Natural selection of the fittest organisms is always accompanied by death of the less fit. The cost of evolution is very high: to generate the few survivors with advantageous traits, many many organisms must die.

At some stage during the early evolution of organisms, the RNA genetic information was replaced by DNA, with the same genetic code.

It is possible that this was related to the need to have greater accuracy in the replication of genes. In present day organisms, DNA is subject to both efficient proof reading to remove initial errors, and also repair of any abnormal features (see Chapter 3). The replication of RNA is about 10,000 times less accurate, which we see in viruses with RNA genomes, such as those causing influenza and AIDS. These viruses are continually producing new mutations, which is one reason they are so hard to combat.

As a result of greatly increased accuracy, present day cells produce offspring which are nearly always error-free and viable, in sharp contrast to early primitive organisms. It is interesting to consider how "optimal" levels of accuracy were reached, because the elimination of errors depends on the utilisation of energy. There must be some point where it becomes counter-productive to eliminate every mistake or error, so this becomes yet another example of the law of diminishing returns. At some point, it also becomes disadvantageous to eliminate every error or mutation, not only because increased metabolic resources are required, but also because our organisms which cannot produce new genetic variation is unable to adapt to changing environments. The importance of genetic variation is readily seen in bacteria today, where rare mutants can become resistant to an antibiotic produced by some other microorganism, or can utilise a new energy source.

The evolution of multicellular organisms must also have depended on accuracy in gene and cell replication. Obviously, such an organism would not be properly organised if a proportion of its cells were not viable. Indeed, we know that there are simple animals today – the roundworm or nematodes – which consist of a constant number of cells in the adult. There is no room for random cell deaths, but it is interesting that selected cells are eliminated by a suicide mechanism, which is one essential part of the overall programme for development from the fertilised egg.

The separation of germ line cells from body, or somatic cells, was a vital feature of the evolution of multicellular organisms. The germ line cells must remain in a fully juvenile state, free of errors or defects, and ready to initiate the developmental programme when the egg is fertilised. The situation for somatic cells is quite different, since they will never be transmitted to the next generation. Their function is to provide the vehicle, animal or plant, which facilitates the transmission of germ line cells. This has resulted in the evolution of a vast array of animals and plants, each exquisitely adapted to a particular environmental niche. Although the germ cells may have special properties, they are very often produced in great excess so only a small proportion may be successful in producing the next generation. Again, we see the continuation of life associated with the death of many germ cells, or of offspring which never survive long enough to themselves reproduce. Sometimes the mortality rate is so high, that the number of individuals of a given species gradually declines and eventually becomes extinct. In these cases, the potential immortality of the germ line is not realised. We know that extinction of species is a regular feature of evolution, during which better adapted animals or plants replace those which are less well adapted.

Animals can afford to relax the accuracy and maintenance in somatic cells, if this increases the resources devoted to reproduction. The set of devices, maintenance or repair mechanisms which are so essential for germ line cells, become uncoupled from those in somatic cells to a smaller or greater extent in different animal species. So the continual dialogue between life and death that occurred in early evolution is also seen as the most advanced products of evolution. Apart from a few simple forms, all animals have bodies which do not survive. They age because in natural environments it is simply counter-productive to try to preserve the complex organisation of cells and structures that characterise these many species. The ability of the body to preserve order indefinitely has been dispensed with. The selfish genes dictate that their progenitors are eventually discarded. The evolutionary forces which were responsible for the origin of life, are also responsible for the senescence and death of the most advanced forms of life. The physical law which states that molecules will adopt a state of maximum disorder, was broken when life first emerged, but it is reasserted when animals age and die. Their highly organised structures break down and the constituent molecules either decay to maximum disorder, or are consumed en route by other organisms. The relationships between asexual and sexual reproduction, germ cells and somatic cells ,ageing and extinction are summarised in Figure 8.

It is clear that molecular defects in large molecules contribute to ageing. Some are mutations in DNA, others are errors in RNA and proteins. It was suggested many years ago that some of these errors may contribute to further errors in protein synthesis, and then further errors. If this happened the machinery for making proteins might become unstable, and this would also be irreversible, so cells would die. This "error catastrophe" theory of ageing stimulated many experiments,

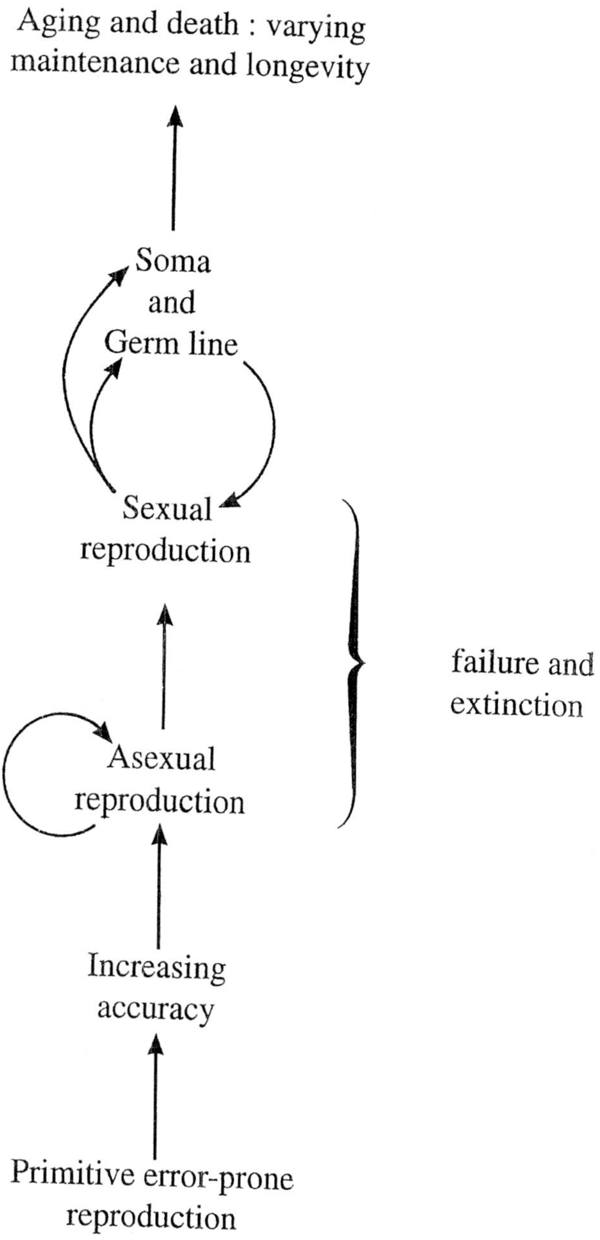

Aging and death : varying
maintenance and longevity

Soma
and
Germ line

Sexual
reproduction

failure and
extinction

Asexual
reproduction

Increasing
accuracy

Primitive error-prone
reproduction

*Figure 8.* Major evolutionary trends. Contemporary organisms reproduce by either
sexual reproduction or asexual reproduction, or both. Ageing of the soma, or body,
is seen predominantly in organisms that reproduce sexually. Germ line cells are
potentially immortal, but their loss leads to species' extinction

and many believe it has been disproved. In fact, the experiments are difficult to carry out, and not enough information has been obtained. The prediction is that human cells would generate significantly fewer errors in proteins than mouse cells. It is significant that we do not even have this information. It is an important gap in our knowledge, because in the evolution of multicellular organisms, the partition between germ line cells and somatic, and the existence of animals with very different longevities, means that there must be optimum levels of accuracy in these different biological contexts. It is not, however, a fashionable field, and very little research is being done.

# Chapter 15. The Road to Discovery

After Peter Medawar delivered his inaugural lecture at University College, London, in 1951, he appointed Alex Comfort to study ageing in his laboratory. Comfort was a polymath: clinician, biochemist, novelist, poet and freethinker, who had only recently become aware of the importance of the study of ageing. As a prelude to his research he decided to survey the whole field and published an important book *The Biology of Senescence* (1956; 2nd Edition, 1963; 3rd Edition 1979). In the Introduction he wrote:

**In almost any other important biological field than that of senescence, it is possible to present the main theories historically, and to show a steady progression from a large number of speculative, to one or two highly probable, main hypotheses. In the case of senescence this cannot possibly be done.**

He went on to cite a very large number of papers which presented different ideas or theories about ageing. The Russian gerontologist, Zhores Medvedev, claimed that there were at least 200 theories and wrote a review in 1990: "An attempt at a rational classification of theories of ageing." As well as theories, there was the accumulation of much experimental data using different animal species, or their cells. These included rats, mice, fruitflies, nematode worms, tiny rotifers, flatworms, cultured human cells, single-celled animals (such as Paramecium) and even fungi and yeast. Much of this data is descriptive and hard to interpret, for example, many of the documented differences between young and old animals, or young and old cells.

For most of the twentieth century there were two major themes underlying the research. One was that ageing is programmed by some kind of biological clock that measured time, or in some cases, the number of cell divisions before senescence. The other was that ageing was due to "wear and tear," or the accumulation of molecular defects in proteins, DNA, and so on. As we saw in the example of the wearing down of herbivores' teeth (Chapter 2), we now know that both genetic programming and wear and tear are important, and cannot be separated.

In 1990, Caleb Finch published a masssive review of research on ageing, "*Longevity, Senescence and the Genome*," with a text of nearly 700 pages, and about 4000 cited publications. Although it can be regarded as a *tour de force*, it did not come to any firm conclusions about the biological reasons for ageing. For most of the twentieth century, ageing was still a major unsolved problem of biology. Apart from the two major themes just mentioned, there were many more specific theories that were, in effect, competing with each other. Many of those who proposed a theory believed that it alone could explain the totality of ageing, for example, the free radical theory, or the somatic mutation theory. However, experimental tests of a given theory were never definitive, at best, they provided limited support. This is not surprising, because in Chapter 4 we saw that there are multiple causes of ageing, and that there is some truth in all the major theories of ageing.

In most books and reviews of ageing, the data obtained from multiple experimental systems is presented and discussed. Oddly, the one animal that is rarely included is *Homo sapiens*. Yet more is known about the age-associated pathologies in man than in any other species. So the thousands of biomedical scientists studying these pathologies (cardio-vascular disease, dementias, cancer, and so on) are really working on ageing itself, or some would say, the consequences of ageing. There are many excellent text books documenting in great detail all these patho-logical changes. It cannot be denied that when it comes to describing the features of senescence and old age, much more is known about man than any other animal species. This makes it all the more remarkable that this information is largely ignored in books on ageing.

One of the most puzzling features of ageing was the fact that within one taxonomic group, such as the mammals, the same changes occur in aged animals, but at very different rates. A good example is the cross-linking of the protein collagen, an essential component of many tissues, and especially connective tissues The chemistry is the same in rats and humans but the *rates of accumulation* are very different. Other examples are discussed in Chapter 6.

Both Peter Medawar, in 1951, and George Williams, in 1956, discussed the evolution of ageing. Both realised that natural popula-tions of animals are age-structured, because high mortality rates means that there are many more young adults than old ones. In fact, as we have seen, few individuals reach old age. Medawar suggested that because there is little natural selection acting on old animals, delete-rious mutations that act late in life would accumulate. He cited the example of the inherited disease Huntingdon's chorea which manifests itself only in the fourth or fifth decade of life. Williams proposed that there would be selection for individual mutations that had beneficial effects early in life, and deleterious effects late in life. Thus both proposed that early and late-acting mutations exist. However, if ageing itself does not occur at all (as in many plants), then the organism is essentially in a steady state, or ageless, so there is no such thing as "early" and "late" acting genes. Medawar and Williams are pre-supposing that ageing already exists, and it is this that can be modulated by evolution.

A different theory for the origin of ageing became necessary, and especially one that took account of important advances in biology. This was proposed by Tom Kirkwood in 1977, and came to be known as the "disposable soma" theory of the evolution of ageing. Initially this belonged to the "wear and tear" category, because it was based on the view that ageing was due to the accumulation of errors in large molecules, such as proteins and DNA. It was known at this time that there are "proof-reading" mechanisms that ensure that not too many errors appear in such molecules. It was also known that such accuracy-promoting mechanisms depend on the expenditure of energy. Thus, to maintain the integrity of the body, there must be investment of resources. However, there is little point in preserving the body in a natural environment in which most animals die from causes other than ageing. So there must be some optimum which ensures that an animal survives long enough to produce offspring, but does not survive indefinitely. The theory proposes that there is a trade-off between reproduction and preservation of the body, or soma, hence the disposable soma theory of ageing. As we saw in earlier Chapters, this neatly explains why mammals that develop and breed rapidly are short lived, and those that develop and breed slowly are long lived. The former group invest a greater proportion of its total resources in reproduction, whereas the latter invest more in preserving the body.

It soon became apparent that the maintenance of the body does not only depend on the accuracy of sythesis of large molecules. There are many other maintenance mechanisms, as described in Chapter 3. So it became obvious that the totality of maintenance depends on a substantial proportion of all the energy resources available to an animal. When scientists started to compare maintenance mechanisms in animals with different life spans, it became clear that they were more effective or efficient in long-lived ones than shortlived ones, In other words, there is a direct correlation between maintenance and longevity, as described in Chapter 6.

One of the most puzzling features of ageing had been solved. **The same changes in cells, tissues and organs occur during the ageing of short-lived and long-lived animals, but at very different rates. This rate depends on the efficiency of maintenance**. Everything now begins to fall into place. Age-associated pathologies are due to the eventual failure of maintenance. These pathologies are studied by biomedical scientists who are specialists, and concerned only with only one of the many of the age-related diseases. They do not have a broad view of ageing itself, but only one part of it. There is also a huge amount of information available about the many different maintenance mechanisms. Most of this information has been been obtained by scientists who do not work on ageing, and indeed may have little interest in it.

From all this research, several very important conclusions can be drawn. First, there are multiple causes of ageing, as outlined in Chapter 4. Second, there are multiple body maintenance mechanisms that together consume considerable resouces. Third, there is not one true theory of ageing, but instead most of the more important ones have a significant degree of truth. Fourth, the number of scientists studying age-associated disease in man, and also on maintenance mechanisms, is far greater that the number of scientists whose stated discipline is ageing itself. Fifth, the whole field of ageing encompasses a huge portion of of biology. To properly understand ageing we must not only haver a broad view, but also extensive biological and biomedical knowledge. Sixth, maintenance mechanisms are very complicated and depend on the activity of large numbers of genes. Therefore, hundreds if not thousands of genes influence in one way or another all the processes of ageing. Seventh, the evolution of the lifespan of any given species occurs over very long periods of time. In this evolution, whether to shorter or to longer lifespan, the various causes of ageing become synchronised. The allocation of resources to maintenance gradually changes in one direction or the other.

**The final major conclusion is that at the end of the twentieth century ageing was no longer an unsolved problem in biology. The so-called "mystery" of ageing had been solved. For most of that century, scientists working on ageing could not see the forest for the trees, because they were all studying individual trees.**

Some discoveries in science are not recognised for a very long time, and two of these were mentioned in the Preface. In the case of ageing, there is not much doubt that the insights and understanding recently obtained, will not be recognised for a very long time. So much that is written about ageing is misinformed or pseudo-scientific, or both. There will be many new theories of ageing proposed, probably with the assumption that all the others are incorrect. There is a widespread view that rejuvenating procedures will greatly extend the lifespan in the future. There is both the gullible public, and the gullible media. With time this will have to change, and the information and understanding we now have about the biology of ageing will be accepted.

# Chapter 16. Resolution of the Paradox

At the outset I raised several related questions. Why do animals have finite lifespans? Why do different mammalian species have lifespans varying by a factor of thirty or more? Why do none live indefinitely? To which can be added, why do we live as long as we do? The previous chapters provided direct or indirect answers to these questions.

The major paradox is an evolutionary one. It is not difficult to imagine a planet inhabited by living organisms which do not age. For example, we could envisage the emergence of unsophisticated forms of life which completely fill the available environment. These organisms might just replace any dead or dying components with new growth. There would be, in effect, a steady state in which all available energy is used for the continuation of life, and all appropriate environments are fully colonised by such life. Order, of a sort, would have triumphed over the physical law which states that all matter reverts to a state of maximum disorder. In somewhat more specific terms, we could imagine a plant biomass, perhaps with rather few species, continually propagating itself without ageing, drawing energy from sunlight and utilising inorganic components for growth. For all we know, there may be existing planets in the universe which harbour life such as this. Why did this not happen on planet earth? Why did life not evolve into such a terminal and uninteresting state?

Perhaps the primary answer comes from a consideration of the marine environment. It is often said that water is essential for the emergence of life, but not that oceans are necessary for the complexity of living forms. Certainly the seas could also support amorphous plant growth, but only at the surface where light penetrates. This would provide the opportunity for organisms which obtain their energy from plants, namely, marine animals. These animals may originally have had simple body plans, but they would have the ability to move around to find their food. With competition between such animals, more and more sophisticated forms would evolve, competing with each other, and increasing their numbers. As plant food became depleted, there would have been evolutionary pressure to seek new food in the non-marine environment, which we can assume had by then been fully colonised by plant life. Several animal forms may have invaded this new inhabitat at different times, but in terms of biodiversity the results were eventually dramatic. To be successful, land-based animals must be efficient in obtaining energy from plants, in reproducing themselves, and in warding off animal competitors. A successful life style was accompanied by many variations in body plan, including bodies with non-dividing cells, as we see in roundworms and many insects today. These organisms efficiently transmitted their genes by sexual reproduction, but did not preserve their body or soma. Ageing had evolved in land animals, which is not to say it did not evolve in aquatic environments as well, for similar reasons. Indeed, it is likely that ageing evolved separately in several different taxonomic groups.

The colonisation of land by plants and animals inevitably lead to the huge biodiversity we see today, namely, innumerable plant species with different morphologies, life styles and ecological niches, and animals which feed on plants, or on each other, adapting to an ever-increasing range of habitats. Vertebrates become the largest land and marine animals, and their body plans are incompatible with indefinite survival. Even the longest-lived, such as whales or giant tortoises, survive only a minute fraction of evolutionary time. The warm blooded vertebrates, mammals and birds, eventually become the most successful land animals, and in both groups adults retain a constant body size, and exhibit the characteristic features of ageing. Humans age because they evolved from species that already had an ageing mechanism, or set of mechanisms, built into their body plans. Interestingly, humans evolved a greater longevity because they became the most successful in colonising the environment (as outlined in Chapter 8).

All this is part and parcel of evolution by natural selection. In the long run, it benefits animals to have a greater reproductive potential than would seem necessary to replenish the population. Darwin and Wallace independently realised that the excess in animal numbers lead to competition between individuals, and provided genetic variation existed, the most successful would be selected. This provides the driving force for greater evolutionary success, but the measure of success really depends on the animals ecological niche, or actual habitat. The elephants grazing in the forests of Africa or India represent one kind of successful life style, in which the animals are large, breed slowly, suffer low natural mortality and have long lifespans. Another kind of success in the forest undergrowth is demonstrated by small rodents which are small, breed rapidly, suffer high natural mortality, and have short lifespans. The longevity of each animal species can ultimately be related to its ecological life-style. However, the longevity is set by the extent of its resources which are allocated to maintenance of the body, which in turn depends on the trade-off between reproduction and maintenance.

If success is equated with the probability of an individualís survival in a natural environment, then it follows that those species which have the lowest annual mortality rates will survive longest. So we see the largest herbivores, higher primates and largest marine mammals sometimes surviving in the wild for long periods. In these circumstances, their ageing may influence reproductive success as for example, when a dominant male Walrus or stag eventually loses his harem of females to a younger and stronger male. In such situations, it can be shown that there will be selection for increased longevity. The selection may be on males, but the females have almost the same complement of genes. In hominids, selection for longevity operated primarily on later reproduction in females.

Whatever the longevity of a given species, very similar processes lead to senescence, but they operate at different rates. One of the main causes of ageing is the inability of the organism to replace cells in vital organs, such as the heart or brain. Individual cells die either through the accumulation of genetic damage in their genes and chromosomes, or through the inability to get rid of defective proteins, or by breaking down such proteins to harmful smaller fragments. The relative importance of genetic or protein changes is a matter for future research. What we can say now that the ability of each cell to maintain its viability has defined limits. Human neurons or heart cells are able to survive very much longer than those of a mouse or rat. From many published experimental investigations we now know that the efficiency of important maintenance mechanisms is related to the animals maximum lifespan. We need to know a great deal more about the reasons for the eventual failure of the maintenance of non-replaceable parts of the body. This will help us understand the reasons for the outset of a whole range of age-associated diseases. These diseases consume a disproportionate amount of medical care in developed countries. It is a far better strategy to delay or prevent onset of these diseases, than to merely treat each one as it arises. Such treatment can be extremely expensive, involving a variety of life-support systems, and it is counter productive, if successful treatment is soon followed by the emergence of a different medical problem in the very same individual.

It is usually taken for granted that the replacement of defective parts of the body is beneficial. False teeth have been available for decades, as has skin grafting and plastic surgery. More recently, there have been very successful kidney transplants, as well as heart, lung and liver transplants. All these patients need life-long suppression of the immune response to prevent the rejection of foreign tissue, and this suppression may a variety of side effects, including the development of tumours. Because there is a chronic shortage of human donors, there is increasing discussion of xenotransplants, that is, the use of organs from animals. This approach poses even greater problems of rejection of the transplanted organ by the recipient. Would it not be better to use non-living components wherever possible to replace defective parts of the body? Plastic or metal "biomaterials" are increasingly being used in a variety of surgical procedures, and it is not difficult to imagine the development of more complex technology, for example, an efficient pump to replace the heart.

One basic problem is the spread of the belief that life-extension will be possible in the future. The belief takes many forms, and it is unfortunate that they become inextricably linked with future scientific advances. Some cling to the view that extreme longevity, or even immortality, will be achieved. Individuals pay good money to have their deceased bodies frozen in liquid nitrogen, in the hope that eventually resuscitation and continued life will be possible. Much more common is the view that scientists will find ways and means of preventing senescence and ageing. For all the biological reasons that have been outlined, this is totally unrealistic. Nevertheless the anti-ageing medicine lobby has become very strong. It does not take much notice of what is known about ageing. Instead, it pits its faith in rejuvenation, regeneration, stem cell technology. It make extravagent claims about life extension in *the near future,* and it is breathtakingly arrogant, because it is in effect saying that it can do better than all those thousands of biomedical scientists who are trying prevent or cure all the age-associated diseases briefly described in Chapter 10.

It would be far better to accept the fact of mortality, and to vigorously explore all avenues to increase the healthspan. There is much to be learned about the effects of life style on age-associated disease. At the moment, there is a huge market for life-extension products which bears little relationship to scientific information. For example, the view that anti-oxidants are very beneficial is widespread, but the actual evidence is flimsy or absent. There are now fairly strong evidence linking a diet high in fruit, vegetables and fibre with improved health, notably, a reduced risk of cancer and heart disease. Much more information is needed from epidemiological studies, longitudinal studies, as well as research on all those deleterious changes which give rise to age-associated disease and a shorter than ideal healthspan. It may well be that the eventual outcome of all future research will be some increase in average, or even maximum lifespan. But the ability to modulate longevity is very different from preventing ageing itself.

To what extent can ageing be regarded as the ultimate disease? It has been joked that ageing is a sexually transmitted terminal disease. It has also been asserted that ageing is an entirely natural process and therefore quite different from any disease. The biogerontologist Leonard Hayflick wrote:

**To understand ageing, it is necessary to distinguish between normal ageing and the diseases that are associated with old age. Although the term "normal ageing" is frequently used, it is not a good choice of words because it implies that there is such a thing as abnormal ageing. That, of course, is absurd. No one experiences abnormal ageing. Normal ageing is simply ageing.**

In one sense this is correct, in another it is not. No physician can write on a death certificate that an individual died of "natural ageing"; instead a specific cause, or causes, of death is inserted. However natural senescence may be, it ultimately leads to organ failure and death, and organ failure is not different from one or another age-associated disease. Some of the confusion arises from the fact that very old individuals may be completely free of particular age-associated diseases, such as Alzheimer's disease or cancer. Yet if they lived long enough they probably would ultimately develop such diseases. The issue is really a statistical one: there is a given distribution of lifespans in populations with good healthcare, and there is a given distribution of any particular age-associated disease. The two are not identical, but they do overlap. The probability of exhibiting each particular age-associated disease increases with old age, but it may never develop, and death is then due to some other cause which is disease-related. As we have seen, ageing is multi-causal, but the loss of function of various organ systems is not at all synchronised. It is an unfortunate fact that in many individuals some age-associated disease occurs well in advance of others. Although ageing *per se* may not be a disease, it is certainly due to a cluster of pathological changes in the tissues and organs of the body, as revealed by autopsies carried out on very old people. This brings us back to the healthspan, which is that proportion of lifespan largely or entirely free of disease. A major aim of medical science should be to prolong it, increase the quality of life of old people, and reduce the current amount of help needed by the infirm elderly.

In the past, death was accepted as a common event. It occurred at all ages, usually as a result of infectious disease. Even as late as the last century, childhood deaths were common in large families. They were expected, and were therefore less of a tragedy than is the case today. Instead, death has achieved the status of a taboo in many current societies. Because death is hidden from view, individuals are ill-prepared for it, and come to fear it. Would it not be far better to understand the biological reasons for ageing, and the reality of death in old-age? Many refuse to accept this reality, and instead put their faith in an eternal non-material afterlife. This is their way of avoiding reality. A more positive approach is exemplified by the author Jack London, who wrote:

... **man's chief purpose is to live not to exist; I shall not waste my days trying to prolong them; I shall use my time.**

In contrast, worrying about the fact of ageing provides no benefit. As well as a positive attitude to one's own finite lifespan, there are other benefits which are even more important.

First, individuals transmit their knowledge and experience, gained during their lifetime, to other members of the community. The social evolution of human populations depends on this transmission. It includes the transmission of existing human knowledge, through education in all its forms. It also includes the transmission of new cultural activity, encompassed in books, various art forms, technology and science. Many are successful in such activities (either as individuals or as members of a group), but many more are the recipients who benefit from them. Second, our transient existence also has the positive function of transmitting human genes to the next generation. The germ line is the quintessence of immortality; the product of a billion years of biological evolution, and sustaining all future generations. Natural ageing and death is the price we pay for successful reproduction. But it is not only reproduction which has made human beings the dominant species on earth. It is the combination of social evolution and the transmission of genes which has lead to success. The genes produced the human brain, and the creativity of the brain lead to all those features of living which are uniquely human. Genetic evolution is slow and social evolution is fast, but both depend on transmission from one generation to the next. In the case of social evolution, each generation learns from the previous one, and then adds new information or understanding. It would be unthinkable for progress to be made in the way that it has, if individuals lived indefinitely. Then knowledge and progress would come to a halt; it would be fossilised and static. Could anyone imagine that the energy and ingenuity that springs from youth and adult life, could be continued by those very same individuals long into the future? Then their brains would be cluttered with memories, and experiences, habits and behaviour would become outdated. The emergence of sophisticated societies and cultures simply could not have occurred without the replacement of each generation by the next. This ageing and death should not be taboo, but instead respected as a completely essential part of human existence and all aspects of human life. The very existence of genetic and social transmission should by themselves provide fulfillment, without the need to worry about our brief lives.

# Selected References

1. The four books mentioned in the Preface are:

Hayflick, L. (1994) **How and Why We Age**. Ballantine Books, New York. 2nd Edition 1996.
Holliday, R. (1995) **Understanding Ageing**. Cambridge University Press, Cambridge.
Austad, S.N. (1996) **Why We Age**. John Wiley, New York
Kirkwood, T.B.L. (1999) **The Time of Our Lives**. Weidenfeld and Nicolson, London.

2. The lecture by P.B. Medawar mentioned in the Preface and Chapter 15 was published in 1952 and republished in 1981:

Medawar, P.B. (1952) **An Unsolved Problem in Biology**, Lewis, London. Reprinted (1981) in **The Uniqueness of the Individual**. Dover, New York.

3. The monographs on ageing mentioned in Chapter 15 are:

Comfort, A. (1956) *The Biology of Senescence*. Routledge and Kegan Paul, London,
Comfort, A. (1964) *Ageing: the Biology of Senescence*. Routledge and Kegan Paul, London.
Comfort, A. (1979) *The Biology of Senescence*. Churchill Livingstone, Edinburgh.
Finch. C.E. (1990) *Longevity, Senescence and the Genome*. University of Chicago Press, Chicago

4. Some articles on ageing by the author, largely in non-technical language:

1984. The ageing process is a key problem in biomedical research. *Lancet* **2**, 1387–1389.
1988. Towards a biological understanding of the ageing process. *Perspectives in Biology and Medicine* **32**, 109–123.
1989. Food, reproduction and longevity: Is the extended lifespan of calorie-restricted animals an evolutionary adaptation? *BioEssays*, **10**, 125–127.
1992. The ancient origins and causes of ageing. *News Physiol. Sci.* **7**, 38–40.
1994. The process of ageing: breakdown of body maintenance? *Stress Medicine* **3**, 3–4.
1996. The urgency of research on ageing. *BioEssays* **18**, 89–90.
1996. The evolution of human longevity. *Perspectives in Biology and Medicine* **40**, 100–107.
2001. Ageing and the biochemistry of life. *Trends in Biochem. Sciences* **26**, 68–71.
2001. Human ageing and the origins of religion. *Biogerontology* **2**, 73–77.
2004. The close relationship between biological ageing and age- associated pathologies in humans. *J. Gerontology, Biological Sciences* **59**, 543–546.
2004. The multiple and irreversible causes of ageing. *J. Gerontology, Biological Sciences* **59**, 568–572.
2005. The evolution of human longevity, population pressure and the origins of warfare. *Biogerontology* **6**, 363–368. **6**, 151–156
2006. Ageing is no longer an unsolved problem in biology. *Annals New York Acad Sci.* **1067**, 1–9.

# Glossary

**Accuracy**. The synthesis of DNA is highly accurate, with about one error per $10^8$ (100 million) bases per cell division. This accuracy depends on at least two proof-reading mechanisms, which detect and repair primary errors. The synthesis of RNA and protein is also accurate, but less so than DNA, with an error rate of about $10^{-4}$ (1 in 10,000) per base or amino acid. The proof-reading is less rigorous.

**AGEs** (advanced glycation end-products). Some chemical modifications of proteins are normal and some abnormal. One of the latter is the attachment of a sugar molecule (glycation). Glycated proteins can aggregate into AGEs, which are often not degraded.

**Alzheimer's disease**. An age-associated brain disease causing an ever-increasing loss of memory. It is associated with the appearance of abnormal proteins and peptides. After the age of 65 the incidence doubles every 5 years.

**Amino acids**. The building blocks of proteins. There are twenty of them, and all have at least one positive and one negative charge. They can be joined head to tail, in a sequence determined by DNA, to form a polypeptide chain. Amino acids can be chemically modified after protein synthesis.

**Atherosclerosis**. The development of abnormal structures on the inner walls of arteries, particularly fatty cholesterol-containing plaques. These can block arteries *in situ*, or after detachment.

**ATP**. Adenosine triphosphate. It is generated by respiration and provides energy for innumerable biochemical reactions that categorise metabolism.

**Carnosine**. A dipeptide consisting beta alanine and histidine. It is present in human muscle (0.5% wet weight), the brain and other body locations (approximately 200 grammes per 60 kg. body weight). Beta alanine is an unusual amino acid not found in proteins. Carnosine slows down the ageing of cultured human cells, and also inhibits the abnormal attachment of sugars to protein molecules (glycation).

**Chromosome**. A long stretch of DNA containing many genes. In plants and animals the DNA is bound to proteins called histones. Each human cell (except red blood cells) contain 46 chromosomes, 23 inherited from one parent and 23 from the other.

**Code**. The genetic code is the means whereby the DNA sequence determines the amino acid sequence in proteins. Three units in DNA determine one of the amino acids, hence the triplet code. (Unfortunately with the DNA sequencing of the whole of the human genome, it is often now referred to, incorrectly, as the human genetic code).

**Collagen**. The most prevalent protein in the body. It has a fibrous triple helix structure with great tensile strength, and is an essential component of connective tissue in all its locations. It has been shown that collagen molecules become progressively cross-linked with age.

**Diabetes**. Early onset diabetes is due to the loss of cells in the pancreas that make insulin. Late onset diabetes is due to defective insulin metabolism, and loss of the control of blood sugar levels.

**Diploid**. A diploid cell contains two copies of the genome, one from each parent.

**DNA** (deoxyribonucleic acid). A long linear polymeric molecule contain just four units of genetic information called bases. These are A (adenine), T (thymine), G (guanine) and C (cytosine). DNA is a double helix in which A and T are paired with each other, and G and C are paired with each other.

**DNA modification**. One of the four bases in DNA, cytosine, can be chemically modified to methyl cytosine. DNA methylation can influence gene expression, and can also be inherited.

**DNA polymerases**. A family of enzymes that accurately replicate the genetic material DNA.

**Drosophila** – see **fruit fly**.

**Duplication** Used in two senses: 1) During cell division the DNA and chromosomes are duplicated; 2) A given length of DNA sequence can be duplicated, either in tandem, or in different positions in the genome. During evolution these duplications my diverge to produce genes with different, but overlapping functions.

**Enzyme**. A protein with a specific catalytic property. Almost all the biochemical reactions that characterise metabolism are carried out by enzymes. They commonly consist on one, two or four subunits, and each sub-unit is made up of a linear chain of amino acids (commonly in the range 100–1000).

**Epigenetics**. All those processes necessary for the unfolding of the genetic programme for development of an organism. Also, information superimposed on DNA, which can be heritable (see **DNA modification**).

**Free radicals**. The important free radicals implicated in ageing are the reactive oxygen species (**ROS**). They are short lived highly reactive incomplete molecules, such as superoxide, hydroxyl ions and hydrogen peroxide, produced as a result of respiration, but also during detoxification by liver cells.

**Fruit Fly**. Several species of the genus **Drosophila** have been used in research on ageing. Pioneering genetic studies by Thomas H. Morgan and his school established the chromosomal basis of heredity.

**Gene**. The fundamental unit of inheritance composed of DNA. Many genes code for specific proteins, but also for RNAs with other functions.

**Genome**. The sum of all DNA sequences in a haploid cell.

**Gametes**. The haploid cells (spermatozoa and eggs) that can fuse to form a fertilised egg or zygote.

**Gonads**. The organs, ovary and testis, that produce eggs and spermatozoa (gametes).

**Haploid**. A haploid cell contains one copy of the genome.

**Hayflick limit**. Leonard Hayflick discovered that human fibroblasts have a finite lifespan in culture (commonly 50–60 sequential divisions), Subsequently it was shown that other cell types also have limited growth in culture.

**Helicase**. An enzyme necessary for unwinding the DNA double helix during DNA replication, or in other contexts.

**Homeostasis**. A physiological term used to describe the complex of regulatory mechanisms that enable the body to function normally. Temperature control is a very important homeostatic mechanism.

**Hormesis**. Sequential treatments that induce stress, for example, mild heat shock. Their cumulative effect may be beneficial.

**Hyperglycaemia**. An excess of glucose in the blood, often associated with diabetes. Hypoglycaemia is a low level of glucose in the blood.

**Hypertension**. High blood pressure, one cause of which is the hardening of the walls of arteries.

**Inbred lines**. The successive mating of male and female sibling mice gradually eliminates intrinsic genetic variability. In normal animals, any given gene inherited from one parent is likely to differ slightly in DNA sequence from that inherited from the other parent. After a given number of generations of inbreeding, the genes inherited from each parent are the same.

**Meiosis**. The cell division necessary to produce spermatozoa and eggs. The chromosome number of a diploid cell is reduced to half (i.e. haploid). This is achieved by one round of chromosomal replication and two cell division. Thus, one diploid cell produces four haploid cells. During meiosis the chromosomal DNA is also re-assorted by the process of genetic recombination.

**Mitosis**. Normal cell division, during which the DNA and chromosomes are replicated, and each daughter cell receives one copy of the genetic material.

**Mitochondria**. Small organelles in the cytoplasm of cells. There are many of them in each cell, and all contain a small molecule of DNA. They are the powerhouses of the cell, generating energy from oxygen and glucose, that is, by respiration (see **ATP**).

**Mutation**. A change in the sequence of DNA which is inherited. Mutations may substitute one of the four genetic units for another (see **DNA**). It may add or subtract a unit, or it may change the DNA sequence in a more complex way.

**Nematode**. Free-living small roundworms containing a constant number of body cells (apart from gametes). The species **Caenorhabditis elegans (C.elegans)** is about 1 mm long and has been used extensively in genetic studies, as well as in experiments on ageing. Parasitic species can be much larger.

**Neoteny**. The evolutionary process whereby an adult retains features of an early developmental state of its ancestor. Thus, an adult Axolotl resembles the tadpole of the salamander. Adult humans have a greater anatomical resemblance to young apes than to adult apes.

**Oocyte**. The cell which will develop into an unfertilised egg.

**Peptide**. When the polypeptide chain in a protein is cleaved, peptides are formed. The accumulation of abnormal peptides in the brain is an important cause of Alzheimer's disease.

**Progeria**. A rare inherited human condition that results in premature ageing. Young children appear normal, but they later develop many of the features of elderly people. Their lifespan is commonly 12–14 years.

**Proof-reading**. The mechanisms whereby errors in the synthesis of large molecules, such as DNA, RNA and proteins, are recognised and either removed or corrected. Proof-reading is particularly important in the replication of the genetic material DNA.

**Proteases**. Enzymes that can cleave proteins to form peptides, which are usually further degraded. They are particularly important when they recognise and act on abnormal protein molecules.

**Proteins**. Linear polypeptide chains containing 20 types of amino acid. They are folded into specific three dimensional shapes and have several major roles: 1) Structural functions, and 2) catalytic enzymes, 3) receptors in membranes, 4) attached to or interacting with DNA and RNA. Proteins are often composed one or more types of subunit, each subunit being one polypeptide chain.

**RNA** (ribonucleic acid). It also has four units of information like DNA. Three are the same as in DNA, but the fourth is uracil (U) in place of thymine (T). The sequence information in DNA is transcribed to messenger RNA (mRNA) by a process known as transcription. The mRNA moves to the cytoplasm and its sequence determines the structure of proteins by a process known as translation. RNA has a number of other functions in protein synthesis and in regulation.

**ROS**, see **Free radicals**.

**Telomeres**. The structure at the end of linear chromosomes consisting of short repeating sequences of DNA. Telomeres are maintained through cell division by the enzyme telomerase, and in its absence telomere sequences gradually become shortened.

**Totipotent**. A cell which is capable of producing all other type of cell. Many of which are fifferentaired. Embryonic stem cells are totipotent.

**Werner's syndrome**. A rare inherited human condition which causes premature ageing. Children and young adults appear normal, but many symptoms of premature ageing appear in the third and fourth decade of life, and the average lifespan is about 45 years. Werner's syndrome is caused by a recessive mutation inherited from both parents.

**Zygote**. The diploid cell produced by the fusion of male and female gametes (sperm and egg).

# Index